National Geographic Picture Atlas of

Our Universe

National Geographic Picture Atlas of
Our Universe

by Roy A. Gallant

Published by
the National Geographic
Society

Gilbert M. Grosvenor
President and Chairman of the Board

Michela A. English
Senior Vice President

William R. Gray
*Vice President and
Director, Book Division*

Prepared by
National Geographic Book Service

Charles O. Hyman
Director

Kenneth C. Danforth
Managing Editor

Staff for this Book

Margaret Sedeen
Editor

David M. Seager
Art Director

Linda B. Meyerriecks
Greta Arnold
Illustrations Editors

Constance Brown Boltz
Associate Art Director

Jennifer G. Ackerman
Ross S. Bennett
Seymour L. Fishbein
David F. Robinson
Jonathan B. Tourtellot
Editors and Writers

Melanie Ann Corner
Mary B. Dickinson
Susan C. Eckert
Shirley L. Scott
M. Washburn Swain
Elizabeth S. Wooster
Research

Charlotte Golin
Design Assistant

Editorial Contributions by
Wendy W. Cortesi
Suzanne P. Kane
Elizabeth L. Newhouse
Lyle Rosbotham
Karen Dufort Sligh
Lise Swinson
Anne Withers

Richard S. Wain
Robert C. Firestone
Andrea Crosman
Leslie Adams
Production

Karen F. Edwards
Traffic Manager

Paulette L. Claus
Sandra F. Lotterman
Teresita Cóquia Sison
Editorial Assistants

Clifton M. Brown III
John T. Dunn
David V. Evans
Ronald E. Williamson
Engraving and Printing

John D. Garst, Jr.
Virginia L. Baza
Charles W. Berry
Donald L. Carrick
Gary M. Johnson
Mark Seidler
Alfred L. Zebarth
Publications Art

George I. Burneston III
Index

Art and Maps by
Jay L. Inge
Davis Meltzer
Ron Miller
Ludek Pesek and others

Scientific Consultants

Mark R. Chartrand III
*American Museum-Hayden
Planetarium, Retired*

J. David Bohlin
NASA

Geoffrey A. Briggs
NASA

Owen Gingerich
*Harvard-Smithsonian Center
for Astrophysics*

Noel W. Hinners
Martin Marietta Astronautics

Henning W. Leidecker
Goddard Space Flight Center

Stephen P. Maran
American Astronomical Society

Peter L. Munroe
Educational Consultant

First edition 670,000 copies.
Revised edition (1986) 375,000 copies.
Second printing (1988) 85,000 copies.
Revised edition (1994) 117,000 copies.
Library of Congress CIP data page 284.
343 paintings, drawings, photographs,
and maps.

Foreword

When I was young, I used to look up into the sky and see airplanes. Later, when I flew those airplanes, I realized I still hadn't gotten very high, that I couldn't go wherever I could see. I kept trying and I did get as far as the Moon. To Neil Armstrong, Buzz Aldrin, and me, that seemed a very long trip, but compared to reaching the nearest star it was like moving past a dozen grains of sand on a beach. The Moon is about 385,000 kilometers from Earth, while the nearest star, Proxima Centauri, is 40 trillion kilometers away! Now, back here on Earth with a better appreciation of distances, I borrow the eyes of astronomers to show me those places I will never visit. This beautiful book will help you to do that.

Throughout history, the genius of astronomers has been their ability to let their minds roam to the far reaches of the Universe while their bodies were trapped on Earth. Kepler, tracing the planets in their orbits; Einstein, seeing the Universe emerge from his chalkboard—they have freed our minds if not our bodies.

And perhaps our bodies will not be far behind. Rockets have propelled astronauts to the Moon and fired Pioneer spacecraft completely out of the Solar System. It is now technically possible for people to visit Mars, approximately nine months away. The other planets are almost within our grasp and the moons of Jupiter and Saturn may be the unrealized gems of our Solar System. The cameras of the Voyager spacecraft have transformed Jupiter's moons from mere pinpoints of light in a telescope into intriguing, colorful spheres. Io... Europa ... Ganymede ... Callisto. Even the sound of their names excites me. More enticing still is the prospect of learning even more about Saturn. We already know that one of Saturn's moons, Titan, is surrounded by a thick atmosphere. It is my favorite candidate for human exploration.

Beyond Pluto, we have to admit that the distances seem to be too much for us, at least if we regard the velocity of light as the universal speed limit. However, not too many years ago, experts believed we would never break the sound "barrier." Right now, Einstein's theories seem to deny us the stars. Perhaps someday we can discover how to disembody humans in one place and recreate them elsewhere, to circumvent Einstein's barrier, and to roam the Universe, seeking our peers or our superiors.

The more we see of other planets, the better this one looks. When I traveled to the Moon, it wasn't my proximity to that battered rockpile I remember so vividly, but rather what I saw when I looked back at my fragile home—a glistening, inviting beacon, delicate blue and white, a tiny outpost suspended in the black infinity. Earth is to be treasured and nurtured, something precious that *must* endure.

The next two decades should be the most productive years astronomers have ever known. With space telescopes we will accumulate more information about the Universe than we have since humans began to study the heavens. May we be intelligent enough to comprehend it and sensible enough to use it to solve some of the pressing problems of our unique home.

Michael Collins

A spectacular night launch of Apollo 17 crowns the mission series that landed 12 U. S. astronauts on the Moon.

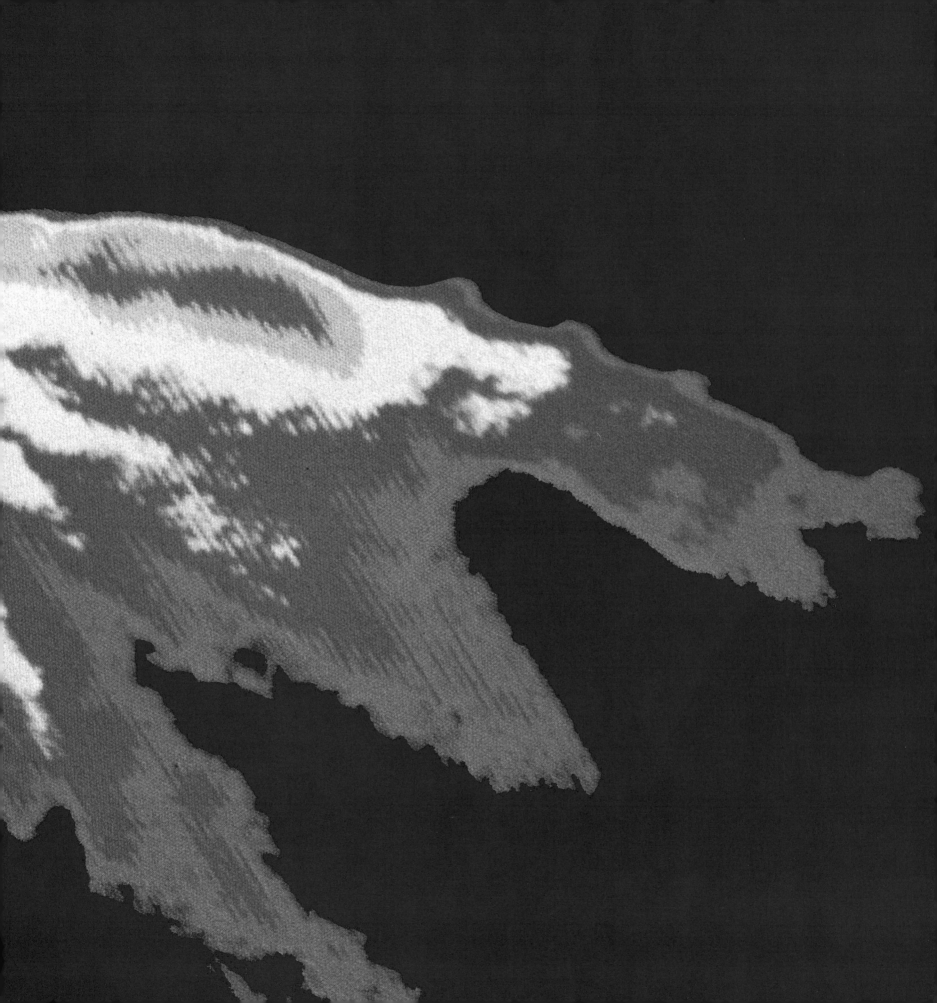

Beginnings

An Egyptian myth about 4,000 years old tells us that a sacred boat carried the Sun god Ra. Each morning the boat rose over the eastern horizon and slowly sailed across the heavenly ocean. At night the boat sank below the western horizon and continued its journey through the underworld until once again it rose in the east and began a new day.

When ancient Egyptians, harvesting wheat near the Nile River, looked at the Sun, they may have pictured the god Ra— or Re, as he is often called—in his boat. The sky has always been a source of wonder. Ancient peoples the world over knew almost nothing about what the stars and planets were, or how they moved through space. The sky seemed to be the home of gods, demons, and spirits, and sky events seemed to be the workings of the gods. Unusual occurrences—like eclipses, comets, meteors—often brought great fear to people in ancient times.

The Egyptian mythmakers wove a creation story that was linked to the annual flooding of the Nile. In the beginning, says the myth, there was nothing but a vast world-ocean called Nun. The Sun god, then known as Atum, created himself out of the waters. At first Atum made a small mound

The "world tree" of the Norse connected sky, Earth, and underworld. From the top, an eagle surveys the world. Below, an evil serpent gnaws on the tree roots. Mimir, a giant, guards the fountain of wisdom. The three Fates—past, present, and future—braid the thread of life.

Brahma, says Hindu tradition, created the Universe from a golden egg. First he made the waters and in them put a seed. It grew into an egg, which Brahma split open. From the golden half came the heavens, from the silver half, Earth. From the egg came all creation.

of earth to stand on and it became the land. Next, he created lesser gods, one to rule over moisture and another to rule over the atmosphere.

Each year from July to October the great Nile River overflowed its banks. When the flood waters lowered, fine muddy soil, called silt, was left heaped along the floodplain. The first appearance of the mounds of silt reminded people of the small mound of earth which they believed Atum had created to stand on. Warmed by the Sun, this rich river-earth nurtured many kinds of life, both plant and animal. No wonder the ancient Egyptians looked on the Sun as a creator-god.

The world tree Yggdrasill

In the beginning there was nothing, a Norse myth written down about A.D. 1200 tells us. "There was no sand nor sea, nor soothing waves / No earth anywhere, nor upper heaven / A gaping chasm and grass nowhere." Then the gods Odin and Thor shaped the world. Earth was flat and in the center grew the great tree of life, Yggdrasill. The ash tree was watered by three magic springs that never ran dry, and the foliage was always thick and green.

Patterns in the sky

After many ages of observing the constellations, planets, Sun, and Moon, ancient peoples learned to recognize and predict patterns. For example, they knew when the Moon would next be full, quarter, or crescent. Such knowledge could be used to make calendars. And, in fact, the world's first calendars were based on the phases of

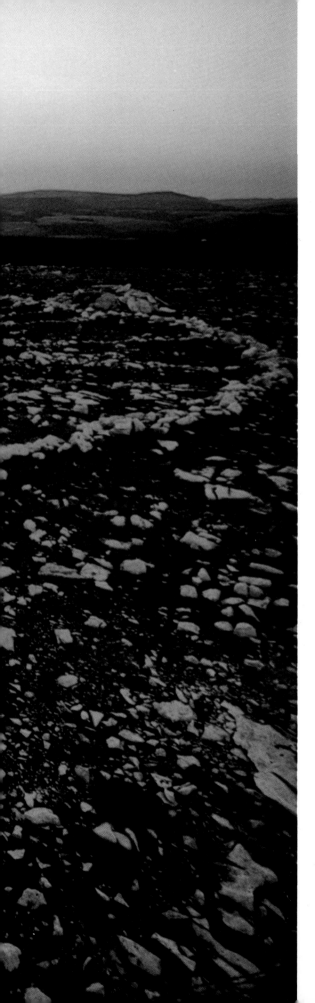

A calendar ring called a medicine wheel told early Indians in Wyoming's Bighorn Mountains that summer had arrived. The rising of the Sun in a line with two piles of rocks, or cairns, marked the summer solstice. According to legend, on this day "the sun is highest and the

growing power of the world the strongest." The 28-spoke wheel is about 24 meters wide, with a central cairn for a hub and six others around the rim. Sighting along different sets of cairns, Indians also observed the summer solstice sunset and the positions of three rising stars (below).

the Moon. Sky watchers noticed that between two full Moons there were sometimes 29 days and other times 30 days. The year could be arranged in 12 "moonths," adding up to about 354 days. For a few years this lunar calendar would work well, but each yearly cycle ended 11 days too soon. So, after eight years, the lunar calendar would be about three months ahead of the seasons. As agriculture became more important, people realized that the seasons related more closely to the movements of the Sun and stars than to the Moon.

The solar calendar

In Egypt, successful crops depended on the Nile River's annual flood. After measuring time in lunar months for thousands of years, the Egyptians developed a more accurate calendar with the help of a star.

Astronomers observed that, shortly before the Nile reached flood stage each year, the star Sirius appeared on the horizon just before dawn. Such careful observations helped them keep the lunar calendar in step by adding extra months. By 2800 B.C., Egypt had established a 365-day year, about a quarter of a day shorter than the actual solar year. This was the first recorded solar calendar.

By watching the Sun's rising point on the horizon each morning, ancient astronomers found that the point shifted each day. In summer it inched its way a bit farther south each morning. In winter it inched its way back toward the north again. June 21 signals the Sun's most northern point — called the *summer solstice.* Some ancient monuments were positioned to pinpoint

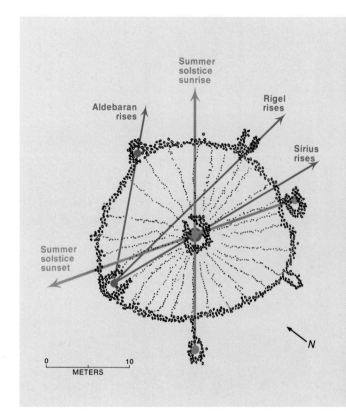

that day: At Stonehenge, England, the Sun rose over a stone marker. Inside the circle of stone blocks, people probably held ceremonies to salute the sunrise and to pray for a season of good crops.

From myths to measurements

Night after night as the stars flowed across the sky, people wondered about those pinpoints of light and asked: What are the stars? Where do they come from? How large are they? How far away? And what is Earth's place in the grand cosmic scheme? Is it really fixed motionless at the center of all creation as it appears to be? In Greece

around 600 B.C., people began to question the old myths about the Universe. Colorful tales of dragons, demons, and Sun gods no longer seemed satisfactory. New thinkers began looking for natural, not supernatural, causes for events in the sky.

The spheres of Eudoxus

One such thinker was Eudoxus, born around 400 B.C. He tried to account for the motions of the stars and planets. To our eyes the stars seem to move as a group across the sky from east to west. But we cannot see them move in relation to each other. So they were called the *fixed stars*.

But certain starlike objects—the planets—move against the background of fixed stars. Stargazers of old counted seven moving objects—the Sun, Moon, Mercury, Venus, Mars, Jupiter, and Saturn—and called them *wanderers*. Usually the planets moved from west to east among the background stars. But then, one of these wanderers would seem to slow down, stop, and mysteriously reverse its direction, or *retrograde*, for a few weeks. (See page 21.) Then, just as mysteriously, it would resume its eastward course.

Eudoxus had a solution to this puzzle. He supposed that Earth was motionless and located at the center of the Universe. And this is just what our eyes tell us. He imagined that a series of nesting transparent spheres enclosed Earth and rotated about its center. He supposed that all of the fixed stars were attached to the inside of the outermost, and largest, sphere. So people saw them move across the sky as this star-sphere turned on its axis. The next inner-

most set of spheres caused the motion of Saturn against the background of fixed stars. The next caused the movement of Jupiter, the next Mars, the Sun, Venus, and so on. Eudoxus never built a model. His system was only a theory and it was complicated but, except for Mars, it did account to some extent for what people observed as the planets moved against the stars.

Earth's shape, size, and motion

While most people thought that Earth was flat, the great Greek teacher Aristotle reasoned otherwise. He had observed that during an eclipse Earth cast a curved shadow on the Moon. Aristotle also believed that of the four elements—earth, air, fire, and water—the heavy ones, earth and water, pulled together into a spherical shape. In Aristotle's time no one knew Earth's size. Later, Eratosthenes made an amazingly accurate measurement using basic geometry.

Aristarchus was possibly the first astronomer to suppose that Earth was simply one of several planets all revolving about the Sun. Also, he said that the stars seemed to parade nightly across the heavens because Earth rotates on its axis. Scholars of his time refused to accept either idea. Their picture of a Universe with a motionless Earth at the center made too much sense to be abandoned easily.

Around 150 B.C. lived a man who has been called the greatest astronomer of ancient times—Hipparchus. By measuring the size of Earth's shadow cast on the Moon during an eclipse of the Moon, Hipparchus estimated the Moon's distance and size

Eratosthenes, a Greek scholar living in Egypt about 230 B.C., knew that every June 21 sunlight fell directly down a well at Syene (now Aswan). Erecting a pole in Alexandria some 800 km to the north, he measured the angle of the pole's shadow at high noon. Geometry told him that if the well and the pole were extended to meet at the center of Earth, the two lines would form the same angle. Both would be 1/50th of an entire circle. Multiplying 50 times the distance from pole to well, he came close to Earth's circumference of about 40,000 km.

Earth wobbles like a spinning top. This causes its north-south axis to point to different parts of the sky during a 26,000-year cycle called precession. Today the North Pole points to Polaris, which we call the North Star. In 13,000 years Earth's North Star will be Vega.

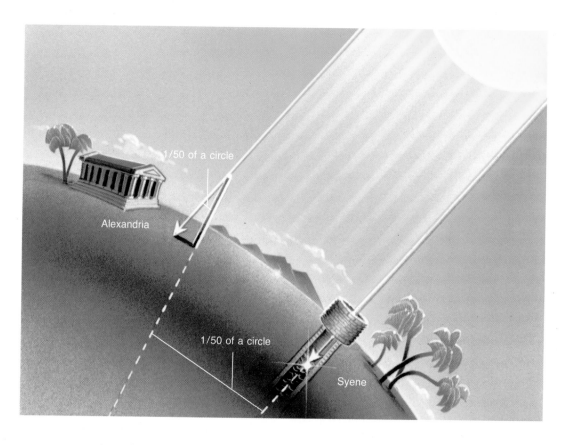

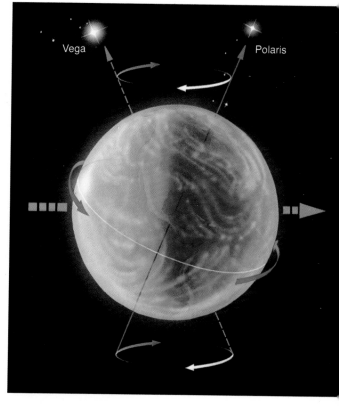

very accurately. He also drew up the most complete star catalog that had ever been made. It showed the brightness and positions of 850 stars. By comparing his observations with those of earlier astronomers, Hipparchus made another important discovery—*precession* (above right).

Ptolemy's epicycles

Greek scholars greatly influenced thinking in the ancient world. Around A.D. 130 the leading astronomer was Claudius Ptolemaeus, called Ptolemy. His views set the climate of astronomical thought for the next 1,500 years. Though he studied the idea of a Sun-centered Universe, Ptolemy supported the old Earth-centered concept. He also rejected the idea that Earth rotates on its axis. If Earth rotated, he reasoned, birds would have their perches whipped out from under them. But Ptolemy did work out a much better scheme to explain the motions of the planets. With a complex arrangement of circles called *epicycles*, he accounted for retrograde motion and for the periodic brightening of the planets as they came nearer to Earth. Ptolemy's system, using numerical data, was the first to make accurate, day-by-day predictions about planetary motion.

The Roman calendar

Roman scholars showed more interest in the superstition of astrology than in astronomy. But Rome did make one significant contribution to the science of astronomy. By 46 B.C., the calendar had fallen hopelessly out of pace with the seasons, and Julius Caesar decreed that the length of the year should be 365 days plus one extra day every four years. But, by the 1500's, people realized that the Julian calendar had been falling behind the seasons at the rate of one day every 125 years. In 1582 Pope Gregory XIII corrected the error. He ordered that three leap years be

Gemini

Taurus

Aries

Pisces

Aquarius

The Roman zodiac, one of several in the ancient world, is used by astrologers today. An imaginary star path through which the Moon and planets traveled, it had a constellation, or sign, for each of its 12 divisions. The Romans believed that the Sun revolved around Earth.

skipped during every 400 years. This Gregorian calendar is the one we use in the Western world today.

The Arabs as preservers

After the time of Ptolemy, much quarreling broke out among the nations surrounding the Mediterranean Sea. In Alexandria, the greatest library of antiquity was destroyed. Fortunately, in the early 800's, there lived an enlightened Arab ruler, Caliph Harun-al-Rashid. He set up a collection of ancient works which later became a research center called the House of Wisdom. Scholars gathered there, in Baghdad, bringing priceless copies of old Greek texts, which they translated into Arabic and so preserved. One of these was Ptolemy's principal work, which they titled *Almagest.* The Arabs had many fine instruments for measuring the positions of the stars and planets. Among them was one called the astrolabe. With this circular instrument and star map, they could sight stars and perform many astronomical calculations. Westerners later used a simplified version to navigate by the Sun and stars. To this day, stars with Arabic names—such as Algol, Zubenelgenubi, Betelgeuse—remind us of the important role the Arabs played in astronomy.

Copernicus and a new revolution

By the 1500's European scholars were rereading the old Greek masters. One such scholar was the Polish astronomer, Nicolaus Copernicus. Over nearly 30 years, Copernicus worked out the details of a Sun-centered, or *heliocentric,* planetary system.

Tycho Brahe, a 16th-century star watcher, points to the heavens in this painting of his observatory, Uraniborg, on the Danish island, Hven. His precision instruments measured the positions of 777 stars and five planets then known. His work marked the beginning of modern astronomy.

This concept, more logical and harmonious than Ptolemy's Earth-centered, or *geocentric,* system, gave the necessary blueprint for progress in astronomy.

Copernicus wrote a detailed mathematical explanation of his theories—the greatest scientific work since Greek times. But he hesitated to publish it for fear of ridicule, and because he knew there were still flaws in the details. However, a young admirer persuaded Copernicus to have the manuscript published. It was called *On the Revolutions of the Heavenly Spheres.* On May 24, 1543, at the age of 70, and only hours after first seeing a complete copy of his printed book, Copernicus died.

Tycho builds an observatory

Accurate scientific observations were needed to test the new theory of Copernicus. The person to produce them was a fiery Dane, Tycho Brahe, often known as Tycho. By the time he was 30, Tycho had earned a reputation in several countries—a reputation for being a gifted astronomer as well as for getting into arguments. As a student, Tycho lost part of his nose in a duel and for the rest of his life wore a false nose made of gold and silver.

In 1576 King Frederick of Denmark gave Tycho the island of Hven, a handsome salary, funds to build a small castle, and the finest instruments to work with. Over the next 20 years, Tycho measured the changing positions of the planets, sometimes using a large wall quadrant that appears in the painting on this page. The telescope had not been invented; Tycho's instruments sighted stars as a rifle sights a

Johann Kepler, a German astronomer using Tycho's observations, found that he could plot Mars' supposedly circular orbit only on an ellipse, or oval. The discovery led to Kepler's First Law: Planets move in elliptical orbits with the Sun at one focus. Construction of an ellipse is shown with two tacks acting as the foci. Kepler's Second Law says: A line from the Sun to a body in orbit sweeps over equal areas in equal time. The shaded areas are equal. Because of the Sun's great mass and gravitational pull, the comet (or any planet) moves fastest nearest the Sun.

target. Tycho watched the *nova*, or "new star," of 1572 and the Great Comet of 1577. His accurate measurements showed that both were more distant than the Moon. Therefore, the comet must have passed through a number of those nesting crystal spheres, by then commonly thought to carry Ptolemy's epicycles. Here was proof that such spheres could not exist! The appearance of the nova proved Aristotle to be wrong when he said that the heavens do not change. Tycho's observations of Mars enabled Johann Kepler to develop his laws of planetary motion.

Kepler frames his laws

In 1600 the 28-year-old Kepler joined Tycho in a new observatory near Prague in Czechoslovakia. The following year Tycho died. Kepler inherited the rich harvest of observations and settled down to discover how the planets move in their orbits. After six years of effort, Kepler worked out his first two laws—among the most important discoveries in the history of science. But one question bothered him very much: What force moves the planets around the Sun? He imagined that the Sun sent out "rays" that whipped around as it rotated and drove the planets in their orbits. However, a better answer to the problem was still 50 years in the future.

Galileo and his telescope

Scholars the world over came to respect Galileo Galilei as a giant of astronomy. He was Italian, born in 1564. He had great fun punching holes in widely held scientific beliefs that had not been tested by

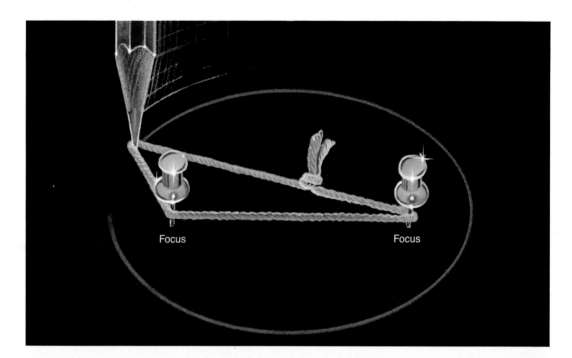

Focus   Focus

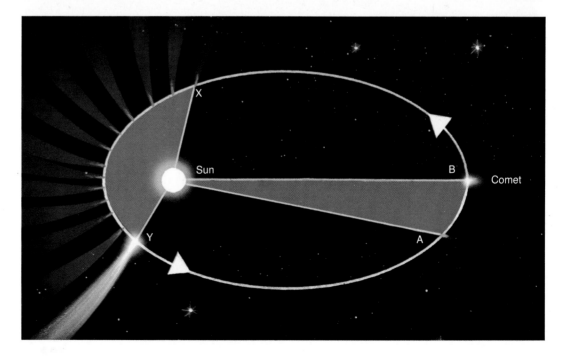

X
Sun   B   Comet
Y   A

A page from Nicolaus Copernicus' book, On the Revolutions of the Heavenly Spheres, *illustrates a theory that brought a new life to astronomy in the 16th century. It shows Earth and the other planets orbiting the Sun. At the time, most people still believed the Sun revolved around Earth. Copernicus also explained why the planets sometimes seem to fall back, or retrograde, in orbit. In the diagram opposite, as the faster moving Earth catches up and passes an outer planet, Mars, the latter appears to observers on Earth to make a loop.*

experiment. For instance, some professors argued that since air does not fall it must be weightless. Galileo compressed air into a leather flask and weighed it. Then he emptied the flask and weighed it again. Since it was lighter than before, he proved that air has weight.

Aristotle had taught that objects fall faster according to their weight. Galileo reasoned that if they were all of the same material, weight wouldn't matter. He tested this by dropping large and small objects of the same material from a high tower. His students, waiting below, saw the objects hit the ground at about the same time.

In 1609 Galileo heard of the invention, in Holland, of a "spyglass," with which distant objects could be seen as though they were nearby. He made himself one of these instruments—a refracting telescope—and began observing the Sun, Moon, and planets. He published a book, *The Starry Messenger,* telling of the wonder of seeing for the first time things which had been invisible to the naked eye. Tradition said that the heavenly bodies were smooth, unblemished spheres of perfection—but Galileo's telescope showed mountains, valleys, and craters on the Moon. When he studied the hazy band known as the Milky Way, he dis-

covered that it was made of countless individual stars. Later, Galileo saw the phases of Venus as positive proof that the Sun formed the center of the planetary system.

Although his discoveries made Galileo known all over Europe, some scholars refused to believe in them. A few refused even to look through the telescope. They still clung to the old ideas of Aristotle and Ptolemy. Church officials warned Galileo not to support the Copernican view that Earth moves around the Sun as the other planets do. For years, he remained silent on this matter.

In 1632 he published a book attacking all arguments against Earth's motion. The Pope turned the matter over to the Roman Inquisition. The next year, Galileo was tried in Rome and forced to deny that Earth moves and that "the Sun is the center of the World." He was kept under house arrest for his few remaining years. By 1637 Galileo had gone blind. But meanwhile, he had written his greatest scientific book—on mechanics and motion. This book is known today as *Two New Sciences.*

Newton and gravitation

The year Galileo died, 1642, Isaac Newton was born in Woolsthorpe, England. Newton solved the problem that had bothered Kepler so much: What force keeps the planets moving around the Sun? And Newton finished Galileo's work on the motion of objects through space.

In 1665 Newton was a student at Cambridge University. That year a deadly plague broke out in London. The University closed and sent everyone home until the

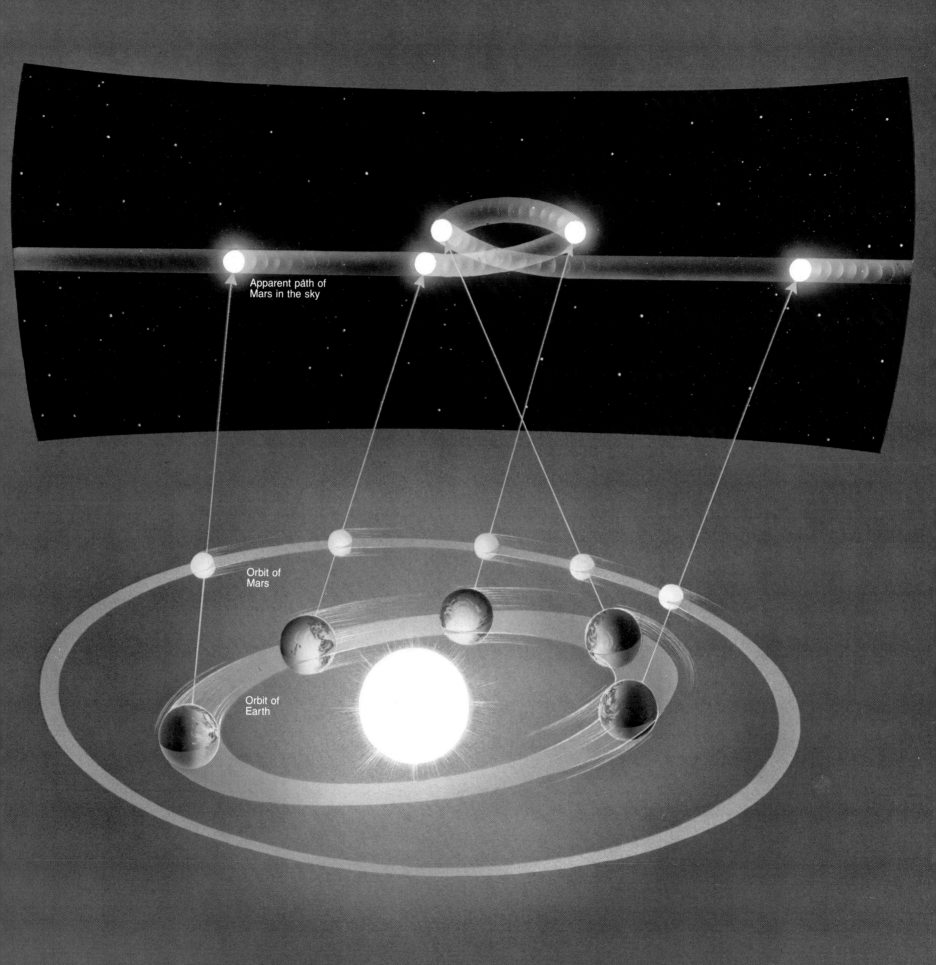

Apparent path of
Mars in the sky

Orbit of
Mars

Orbit of
Earth

Galileo looks through his homemade "optic glass" and draws pictures of the Moon. It is 1609 in Padua, and the Italian astronomer for the first time sees celestial bodies closer than they really are. On what was then thought to be a smooth lunar surface he found mountains and *craters. For many weeks Galileo studied the Moon and sketched its various phases. One of his drawings (below right) appears next to a photograph of a similar phase (below left). Building more powerful telescopes, he observed countless stars never before seen in the skies. He* *saw the four moons of Jupiter and the phases of Venus, and proved that the Moon shines with light reflected from the Sun. In 1612 he saw spots on the Sun. Watching them move across the disk, disappear, then reappear, he concluded that the Sun rotates on its axis.*

epidemic ran its course. As a mathematics student, Newton had become very interested in geometry and the motions of the planets. At age 23, and with a long vacation ahead, Newton set out to investigate the force that seemed to hold the planets captive to the Sun.

He began by wondering why the Moon stayed in orbit around Earth. Why didn't it go flying off into space as a stone whirled around at the end of a string does when the string is let go? Was the force holding the Moon in orbit the same that pulls an object to the ground when it is tossed into the air? Did this force also hold the planets in orbit around the Sun?

Having the idea was one thing. Proving it was another. Newton had to invent a new kind of mathematics—calculus. By the time he returned to Cambridge, he had begun to think about gravitation.

Over the next 20 years, Newton formulated his three laws of motion:

Newton's First Law—An object in motion will keep moving at the same speed and in the same direction unless an outside force acts on it. An object will remain at rest unless a force acts on it, such as the horse pulling the log (page 25).

Newton's Second Law—If a force acts on an object, the object will change velocity— that is, speed up, slow down, or change direction. Change in speed will be in the same direction as the force and proportional to it, but inversely proportional to the body's mass. This means it is easier to move a light body than a heavy one.

Newton's Third Law—For every action, there is an equal and opposite reaction. For

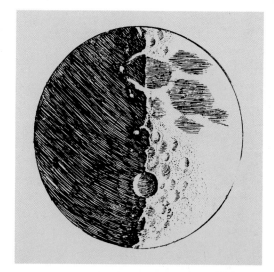

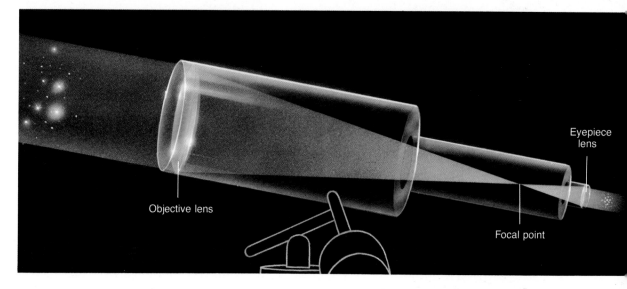

Refractor telescopes used by Galileo and Kepler had two lenses. A convex () objective lens collected the light and bent, or refracted, it to a focal point. The eyepiece lens magnified the image. Galileo placed a concave)(eyepiece in front of the focal point and saw the image *right side up. Kepler used a convex eyepiece behind the focal point (above) for greater magnification. But it captured an inverted image. A third lens reinverted the image but dimmed the light. So it was discarded, and astronomers continue to study the stars "upside down."*

When Isaac Newton saw an apple fall to the ground, he decided that gravitation must also hold the Moon in its orbit. According to his laws of motion, the Moon would continue in a straight line unless Earth's attraction pulled it into a curved path. Similar forces acting on the planets keep them in orbit around the Sun. The horse in the picture (opposite) illustrates Newton's Third Law: When the horse pulls against the log, the log resists. The cannonball (left) cannot escape Earth's gravitational pull unless a big enough blast sends it flying into space.

instance, if you blow up a balloon and let it go, the balloon shoots forward. The forward movement is a reaction equal to the action of the air rushing through the neck of the balloon.

According to Newton's *Law of Universal Gravitation,* any two objects in the Universe attract each other, and, the more massive the objects are, the stronger the attraction. Two elephants would attract each other with greater force than two tennis balls an equal distance apart. The law also says that the closer the two objects are, the more strongly they will attract each other. So Kepler's "mysterious force" holding the Universe together was gravitation. In 1687 Newton's theories of motion and gravitation appeared with the title *Principia*—one of the most important scientific works ever published.

While doing experiments with light and lenses, Newton became interested in telescopes. At the time, there were only refracting telescopes, like Galileo's. One problem with a refractor was that a disturbing rim of color surrounded the circular image. The lens acted like a prism to separate the colors. Newton used a mirror instead of a lens to produce a sharper image. This was the first reflecting telescope (see page 31).

By the time Newton died in 1727, people had come to believe in Copernicus' Sun-centered system. They also accepted the idea that space was infinite and held a vast number of stars. The next big task was to discover the grand plan of the Universe: What was its shape? Where was the Sun's place in it?

Measuring and mapping

The great crystal sphere once thought to hold the stars in fixed places was gone forever. This meant that the stars were free to move about. But did they? No one had seen them move. In 1718 the English astronomer Edmund Halley showed that they do. He compared his own star observations with those made by Greek astronomers about 1,500 years earlier. He was surprised to find that three bright stars, Arcturus, Aldebaran, and Sirius, had changed position.

Herschel gives the Universe a shape

In 1781 a music teacher and amateur astronomer named William Herschel gained fame almost overnight. He had built a fine telescope and with it discovered Uranus. This was the first discovery of a planet since ancient times.

Herschel became a professional astronomer and turned his attention to the motion of stars and the nature of the Milky Way. What arrangement of stars could explain this bright band of light in the sky? In the 1780's Herschel began to count stars systematically, using a method he called *star gauges.* His 18-inch reflecting telescope revealed millions of them!

After examining his data carefully, Herschel came to the conclusion that the Sun was not, in fact, fixed at the center of the Universe. It was moving through space, which made the other stars change their positions. Herschel also reasoned that the Milky Way was shaped like a giant powder puff, with the Sun located close to the center.

Newton sits among objects of his genius: a reflecting telescope and his Principia, *which explained his ideas of gravitation and motion. He holds a prism, which was to become a key to the Universe. In a dark room he placed a prism in front of a sunbeam. The light split into a spectrum.*

Then he turned it back into white light with a lens. Later scientists studied the spectrum and learned about a star's makeup, speed, and temperature. Aboard a NASA training craft known as the Vomit Comet (left), astronauts experience life—and squirting toothpaste—without gravity.

The idea of a galaxy

Herschel's powder-puff picture of the Milky Way remained pretty much unchanged until the early 1900's, when the American astronomer Harlow Shapley examined the sky. Shapley reasoned that the center of our Milky Way system must be toward the constellation Sagittarius, where the sky is most crowded with stars. Shapley had noticed many ball-shaped collections of stars—*globular clusters.* He counted several dozen in the region of Sagittarius but only a few in the opposite direction toward Auriga. He reasoned that globular clusters must be arranged around

the center of our star system. Then he measured the distance to the globular clusters and decided—correctly—that the Sun is located near the edge of the Milky Way.

Around 1924 another American astronomer, Edwin Hubble, discovered that the Universe is much larger than Herschel and Shapley had thought. Here and there among the stars were many dim, hazy patches. What were they? Some proved to be clouds of gas and dust blocking the view of rich star fields beyond. Others, Hubble discovered, were systems of stars like our own. Like Herschel, he called them "island universes." Today we call them *galaxies.*

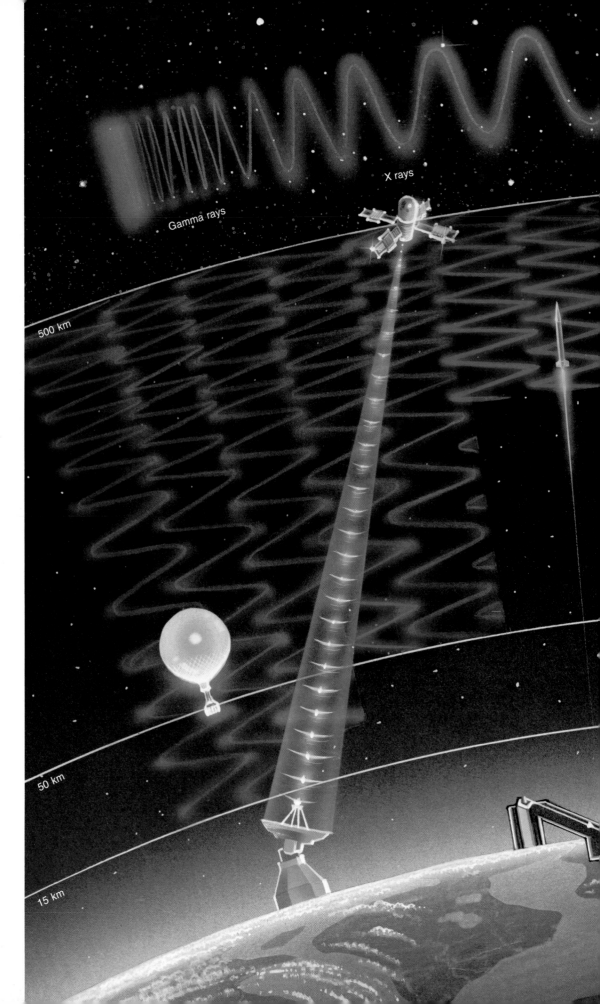

Secrets of the stars shower down on us in the elusive energies of the electromagnetic spectrum. Light and some radio waves reach earthbound receivers. But our atmosphere blocks much of the rest. With planes, balloons, rockets, and satellites, science takes the pulse of space.

Big eyes on the Universe

We learn about the stars and galaxies by studying the light and other energy they radiate into space. Light energy does not need air or any other substance to carry it, contrary to some 19th-century beliefs. Sound-wave energy does. That means radiation can travel through empty space.

To detect light energy, you need no instruments other than your eyes. Light is only one form of radiation sent out by stars and galaxies. Light that appears white to our eyes can be broken up into a rainbow of colors called the *visible spectrum*. We can think of the different colors of light as waves of energy. The waves of any one color differ from the waves of all other colors. Violet light has the most energy and the shortest wavelength, or distance between wave crests. Blue light is a bit less energetic and has a longer wavelength, and so on through green, yellow, orange, and red. Red light has the least energy and the longest wavelength.

The visible spectrum is only a small slice of a larger spectrum of radiation called the *electromagnetic spectrum*. The shortest wavelength and highest energy radiation of the electromagnetic spectrum is *gamma* radiation. This ranges from 50,000 to many million times more energetic than visible light. The second most energetic radiation, with a slightly longer wavelength, are *X rays*. This radiation is 50 to 5,000 times more energetic than visible light. Both gamma rays and X rays can damage the cells of living tissue but in small amounts they have many uses in medicine and science.

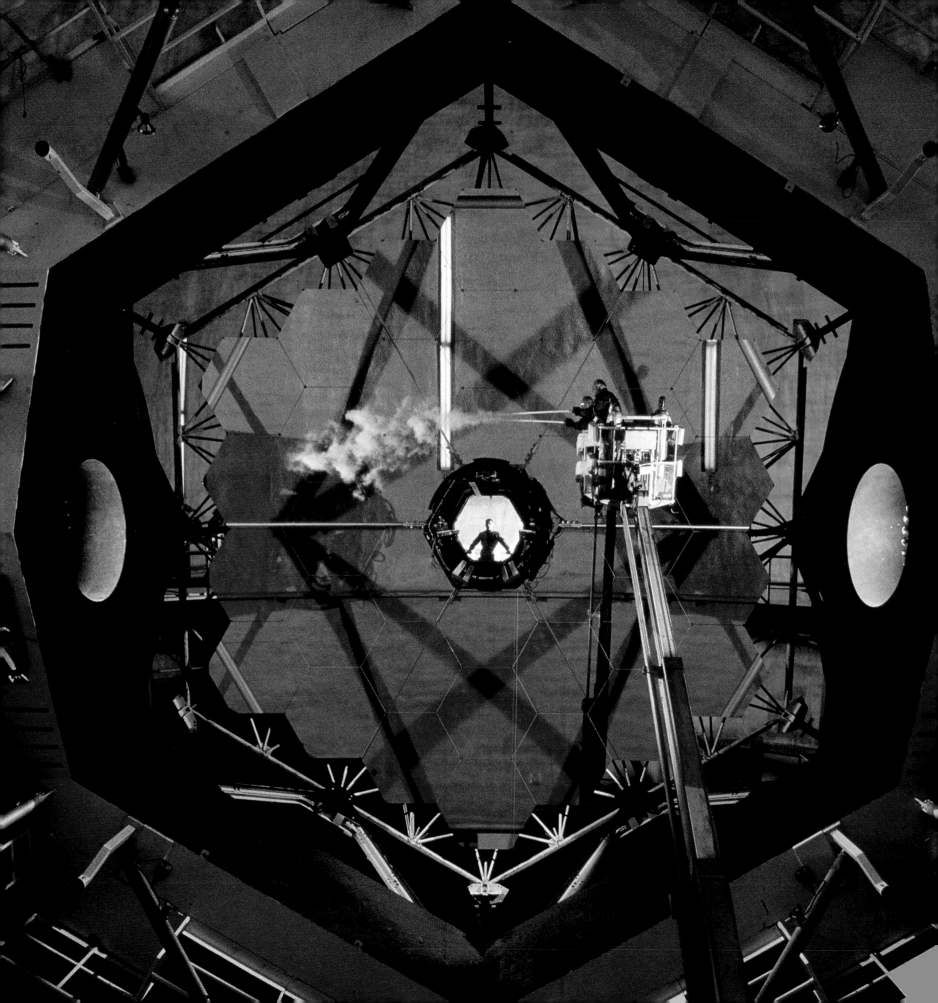

Next to the X-ray region are longer wavelengths called *ultraviolet* (beyond violet) radiation. Then comes the visible spectrum. At wavelengths longer than those of visible light is *infrared* (below red) energy. And beyond the infrared is the least energetic radiation of all, with the longest wavelengths—*radio* energy.

We can think of each energy level along the electromagnetic spectrum as a special window through which we can view the Universe. Our view depends on which energy window we open. (Some examples are shown on pages 66 and 67.)

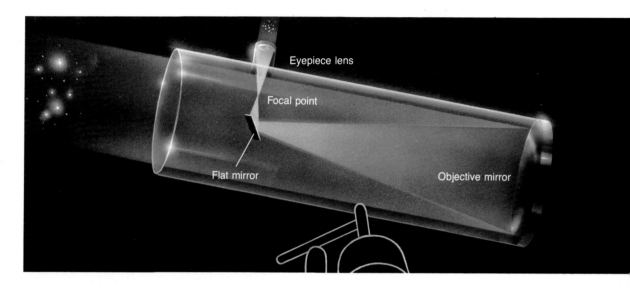

Eyepiece lens

Focal point

Flat mirror

Objective mirror

High-energy astronomy

If Newton and Galileo came back today, they would be amazed by the new telescopes and the new field of high-energy astronomy, which is only about 25 years old. Gamma rays, X rays, and most ultraviolet radiation are blocked by Earth's atmosphere, so the new gamma-ray telescopes, X-ray telescopes, and ultraviolet-ray telescopes must do their observing from space.

In the late 1970's, NASA launched into orbit three telescope-satellites called HEAO (High Energy Astronomical Observatory). HEAO-1 examined the sky through the X-ray window and saw almost five times as many celestial X-ray sources as we had known before. HEAO-2, nicknamed Einstein, carried the largest X-ray telescope ever built. It provided detailed X-ray images of stars, globular clusters, and *supernova* remnants. With its telescopes, HEAO-3 recorded gamma-ray and cosmic-ray emissions from exploding stars. Then, in 1983, the European Space

Agency's versatile X-ray satellite, Exosat, greatly expanded our knowledge of *neutron stars* and other X-ray sources.

High-energy sources like these are targets for X-ray and gamma-ray telescopes. Orbiting ultraviolet telescopes give us information about wind blowing from hot stars, and identify specific atoms and ions in the atmospheres of cooler stars like the Sun. In time these new instruments may help astronomers solve the puzzle of how the Universe is evolving. They also may reveal how clusters of individual galaxies formed and have developed over time.

Optical telescopes

Most optical telescopes are designed to operate from the ground, preferably on a mountaintop above atmospheric haze and far from city lights. But the Hubble Space Telescope, which orbits high above the atmosphere, now gives us our clearest pictures of star clusters and galaxies.

Among the best known optical ground telescopes are the five-meter reflector atop California's Mount Palomar, which is 1,871 meters high, and France's famous Pic du Midi Observatory perched on a Pyrenees peak at a height of 2,865 meters. In the past, Pic du Midi has specialized in lunar and planetary study, but a new two-meter reflector allows its astronomers to study distant galaxies as well.

Not long ago, the world's largest reflecting telescope was the six-meter reflector in southern Russia. But the Keck Telescope, a new reflector on top of Hawaii's Mauna Kea, is nearly double that size at ten meters. The Keck began full operation in 1992. It has four times as much light-gathering power as the Mount Palomar mirror, enough to spot a candle on the Moon. Because it sees twice as far into space—two times farther back in time—it may help answer the question of how the Universe began.

France's Pic du Midi Observatory has been peering through the thin clean air above the cloud-choked Pyrenees for a century. A new telescope, built to produce very sharp images, now studies objects beyond the Milky Way. Others study the Sun, the Moon, and the planets.

Infrared and radio telescopes

Infrared and radio telescopes on Earth give us still other views of parts of the Universe invisible to our eyes. But the atmosphere also limits what these telescopes can detect. For this reason, the year 1983 was an especially exciting one for astronomy. Launched into orbit in January, the Infrared Astronomical Satellite (IRAS) surveyed 98 percent of the sky for 10 months. It returned emissions from more than 200,000 celestial infrared sources, many for the first time. Scientists will be sifting through the returns for years to come.

IRAS was a joint project of the United States, Great Britain, and the Netherlands. It found a sky full of infrared galaxies not detectable by earthbound telescopes. It identified at least five new comets and 10,000 new asteroids. But its most spectacular discovery was a possible planetary system (see page 248) forming around Vega, a star 26 light-years from Earth. Further information about this young planetary system may come from another satellite planned to launch in the year 2002. SIRTF (the Space Infrared Telescope Facility) will be 1,000 times more sensitive to infrared energy than IRAS was.

Radio-wave energy comes from many sources. In 1931 a physicist named Karl Jansky built a large moveable antenna to study radio static. The antenna picked up radio hiss coming from space. Jansky ruled out the Sun as the source and realized that the waves did not come from within the Solar System but from throughout the Milky Way galaxy.

In 1937 Grote Reber, a radio engineer,

As an optical telescope gathers light, so a radio telescope gathers radio waves. Bounced from dish antenna to receiver, they are converted to electronic signals by a vertex box and amplified by a control unit. Computer and display unit produce graphs that show the "noise" of space.

Star trails streak the sky above dish antennas at the Very Large Array in New Mexico. The observatory consists of 27 antennas that move about the desert on special tracks. With computers, astronomers combine the telescope signals to produce radio images of space.

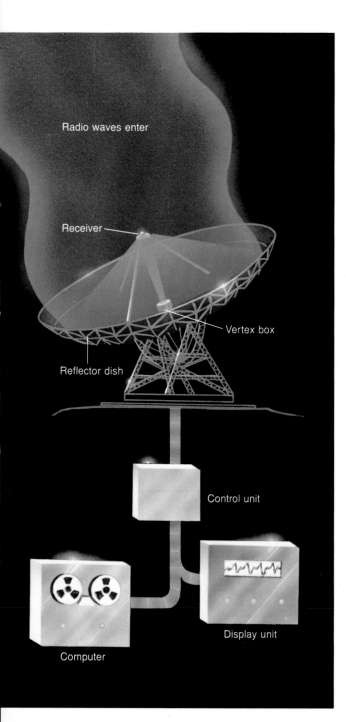

Radio waves enter

Receiver

Vertex box

Reflector dish

Control unit

Computer

Display unit

built the first radio telescope in his yard in Wheaton, Illinois. By studying emissions from different parts of the sky, Reber made the first radio map of the Milky Way. The English scientist J. S. Hey detected radio waves from the Sun in 1942, and 13 years later the American astronomers B. F. Burke and K. L. Franklin discovered that Jupiter sends out radio signals. Since then astronomers have detected radio noise from supernova remnants such as Vela X and the Crab Nebula, from the mysterious *quasars,* from other galaxies, and from the hydrogen gas spread throughout our galaxy.

Radio telescopes give us much information about the Universe that other kinds of telescopes cannot detect. In the late 1960's and 1970's radio telescopes showed that many different chemicals drift through interstellar space. Among them are water vapor, carbon monoxide, ammonia, and formaldehyde, for example. Radio telescopes—by means of radar—also can see through Venus's dense cloud layer and map the planet's surface. These telescopes have become important tools with which to study those parts of our galaxy that are veiled behind clouds of space dust or located in the Milky Way's dense central region.

Astronomy's giant mechanical eyes view the Universe through invisible windows and reveal that the Universe is many, many times more complex and grander than we humans can detect with our unaided senses. If we ever manage to communicate with intelligent beings beyond Earth, it almost certainly will be through one of those windows on the Universe.

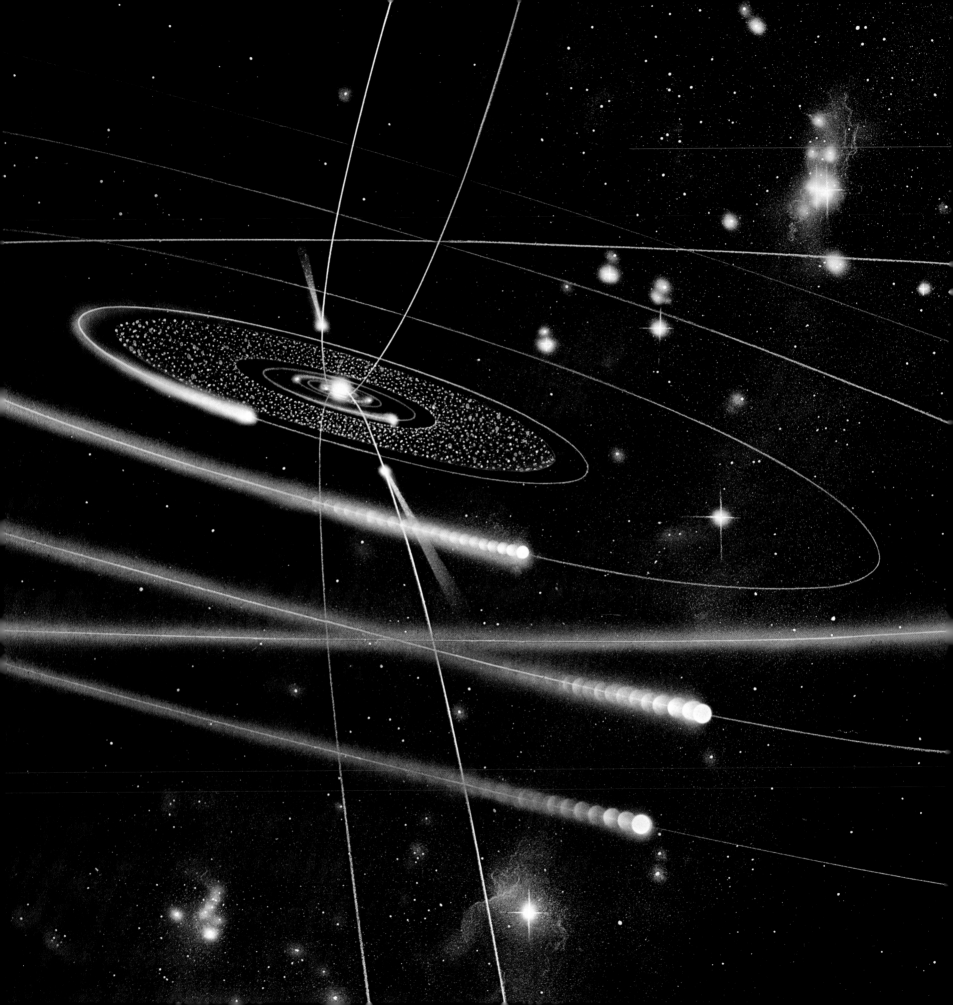

The Sun & Planets

Our Solar System is an orderly community of one star, nine planets, more than 50 moons, thousands of miniature planets called asteroids, and billions of meteoroids and comets.

This sweeping view is from beyond Pluto. The Sun, some 6 billion kilometers away, contains 99 percent of the matter in the Solar System. The inner planets—Mercury, Venus, Earth, Mars—orbit close to the massive Sun. The asteroid belt of rock-metal fragments separates these planets from the outer—Jupiter, Saturn, Uranus, Neptune, and Pluto. All but Pluto and Mercury move in nearly the same plane, unlike most comets, whose long, steeply tilted orbits loop around the Sun.

Stars radiate light of their own but planets shine by the light they reflect from the Sun. They all orbit in the same direction – the closer to the Sun the faster they go. Except for Venus, Pluto, and Uranus, the planets also spin, or rotate, the same way—counterclockwise in this view.

Between the planets the "solar wind" blows. Planetary dust and debris flicker in sunlight and starlight. Speeding meteoroids carry material that may be left over from the birth of the Solar System. But the Solar System is mostly empty space.

The Solar System

What it's mostly made of:

	Number of atoms for every one million hydrogen atoms
Hydrogen	1,000,000
Helium	85,000
Oxygen	661
Carbon	331
Nitrogen	91
Neon	83
Silicon	33
Magnesium	26

The birth of our Solar System remains a puzzle. Some 4.6 billion years ago, scientists think, a vast cloud of gas and dust (A), perhaps shaken by a nearby exploding star, collapsed into a spinning disk (B). Gravitation pulled so much material to the center that pressure and heat there lit a nuclear fire; the Sun began to shine. Other material slowly collected into lumps of hot solids and gases (C), which cooled into planets (D). All are built, like Earth and everything on it, of chemical elements evolved from that original cloud amid the stars—even the oxygen in our lungs, the iron in our blood, the calcium in our bones. And so, in a way, we all are made of stardust.

A About 4.6 billion years ago

B 10 million years later

C 20 million years later

D 120 million years later, and up to today

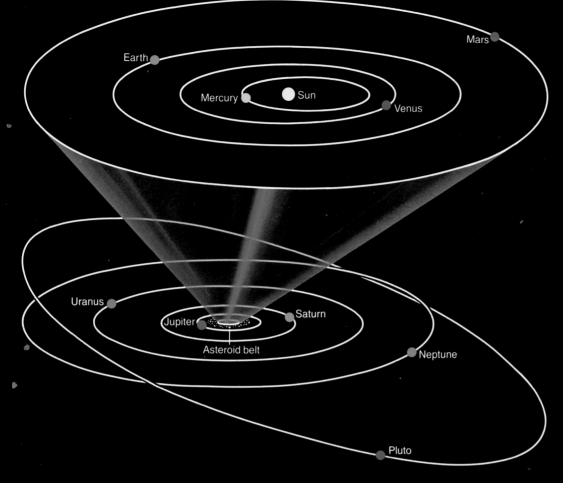

Earth
Mars
Mercury
Sun
Venus
Uranus
Jupiter
Saturn
Asteroid belt
Neptune
Pluto

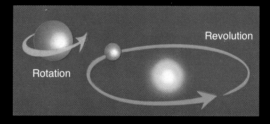

Revolution
Rotation

Speaking in turns—To describe how the planets move, we talk of *rotation* and *revolution* (above). A planet spinning on its axis *rotates*. A single rotation equals one of that planet's days. The planet also orbits around the Sun, or *revolves*. Each revolution equals one planetary year.

This view of our Solar System (above) compares distances. If Pluto's path were about the size of a bicycle tire, the orbits of the four inner planets would fit on the face of a quarter. All planetary orbits seem to form circles, but actually the circles stretch out a little, into *ellipses*.

Orbital tilts—Think of an orbit drawn on a piece of cardboard. The flat cardboard is the *plane of the orbit*. Planetary orbits tilt with respect to each other (see below). To measure the tilts, we use the angles between Earth's orbital plane, called the *ecliptic,* and the other orbital planes.

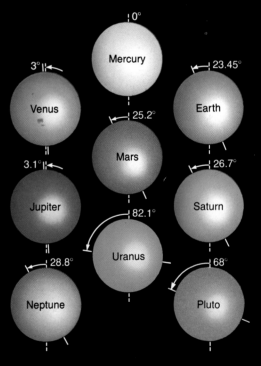

0°
Mercury
3° Venus
23.45° Earth
25.2°
3.1° Jupiter
Mars
26.7° Saturn
82.1°
28.8° Neptune
Uranus
68° Pluto

Like a tipsy top, each planet spins on a leaning axis (above). Only Mercury's axis stands perpendicular to its orbital plane.

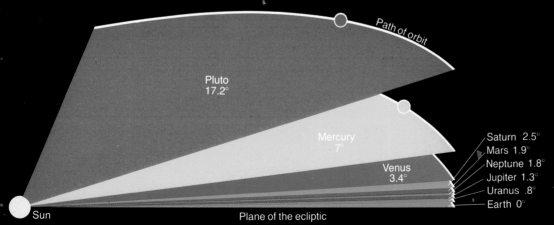

Path of orbit
Pluto 17.2°
Mercury 7°
Venus 3.4°
Saturn 2.5°
Mars 1.9°
Neptune 1.8°
Jupiter 1.3°
Uranus .8°
Earth 0°
Sun
Plane of the ecliptic

A family portrait of the planets, in order by orbit, compares sizes. The inner planets—Mercury, Venus, Earth, Mars—are called *terrestrial*, or Earthlike: small, rocky worlds with metal cores and (except for airless Mercury) shallow atmospheres.

Outward lie the four *gas giants*. No solid surfaces here; deep atmospheres thicken into hot liquid, which reaches all the way to rocky cores. Joining Saturn's famous ring system are the rings of Jupiter and Uranus and the faint, newly found rings of Neptune, which observers long suspected. The last planet, Pluto, is the Solar System's odd man out. Smaller even than Mercury, Pluto seems mostly ice. Satellites orbit every planet except Mercury and Venus. Jupiter and Saturn rule miniature solar systems, totaling five giant moons and many smaller ones.

By studying what our Solar System is like today, astronomers seek to know its past. They ask why some planets are mostly rock and metal, others mostly gas and ice. Did the terrestrial planets begin life as gas giants, only to lose most of their gases to the early Sun's wind and tides? Or did heat and motion near the Sun keep these newborn inner planets from collecting much gas to begin with?

Earth	Jupiter	Saturn	Uranus	Neptune
1 Moon	1 Metis	1 Atlas	1 Cordelia	1 Naiad
	2 Adrastea	2 Prometheus	2 Ophelia	2 Thalassa
Mars	3 Amalthea	3 Pandora	3 Bianca	3 Despina
1 Phobos	4 Thebe	4 Epimetheus	4 Cressida	4 Galatea
2 Deimos	5 Io	5 Janus	5 Desdemona	5 Larissa
	6 Europa	6 Mimas	6 Juliet	6 Proteus
	7 Ganymede	7 Enceladus	7 Portia	7 Triton
	8 Callisto	8 Tethys	8 Rosalind	8 Nereid
	9 Leda	9 Telesto	9 Belinda	
	10 Himalia	10 Calypso	10 Puck	**Pluto**
	11 Lysithea	11 Dione	11 Miranda	1 Charon
	12 Elara	12 Helene	12 Ariel	
	13 Ananke	13 Rhea	13 Umbriel	
	14 Carme	14 Titan	14 Titania	
	15 Pasiphae	15 Hyperion	15 Oberon	
	16 Sinope	16 Iapetus		
		17 Phoebe		
		18 Pan		

Jupiter

Mars

Earth

Venus

Mercury

Limb of the Sun

Saturn

1 2 3 4 5 6 7 8 9 10 11 12 13 14 15 16 17 18

Pluto

Neptune

Uranus

1

11
12
13
14
15

6
7
8
9
10

1
2
3
4
5

1
2
3
4
5
6
7
8

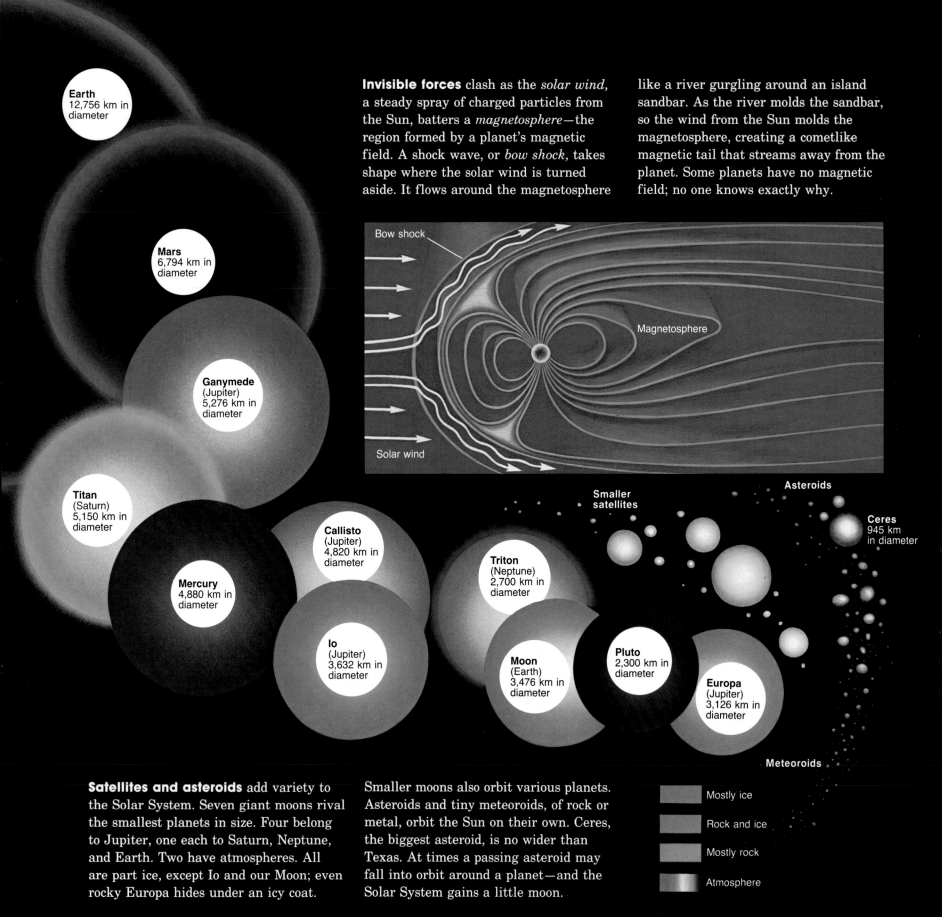

Earth
12,756 km in diameter

Mars
6,794 km in diameter

Ganymede
(Jupiter)
5,276 km in diameter

Titan
(Saturn)
5,150 km in diameter

Mercury
4,880 km in diameter

Callisto
(Jupiter)
4,820 km in diameter

Io
(Jupiter)
3,632 km in diameter

Triton
(Neptune)
2,700 km in diameter

Moon
(Earth)
3,476 km in diameter

Pluto
2,300 km in diameter

Europa
(Jupiter)
3,126 km in diameter

Ceres
945 km in diameter

Smaller satellites

Asteroids

Meteoroids

Invisible forces clash as the *solar wind,* a steady spray of charged particles from the Sun, batters a *magnetosphere*—the region formed by a planet's magnetic field. A shock wave, or *bow shock,* takes shape where the solar wind is turned aside. It flows around the magnetosphere like a river gurgling around an island sandbar. As the river molds the sandbar, so the wind from the Sun molds the magnetosphere, creating a cometlike magnetic tail that streams away from the planet. Some planets have no magnetic field; no one knows exactly why.

Bow shock

Magnetosphere

Solar wind

Satellites and asteroids add variety to the Solar System. Seven giant moons rival the smallest planets in size. Four belong to Jupiter, one each to Saturn, Neptune, and Earth. Two have atmospheres. All are part ice, except Io and our Moon; even rocky Europa hides under an icy coat.

Smaller moons also orbit various planets. Asteroids and tiny meteoroids, of rock or metal, orbit the Sun on their own. Ceres, the biggest asteroid, is no wider than Texas. At times a passing asteroid may fall into orbit around a planet—and the Solar System gains a little moon.

Mostly ice

Rock and ice

Mostly rock

Atmosphere

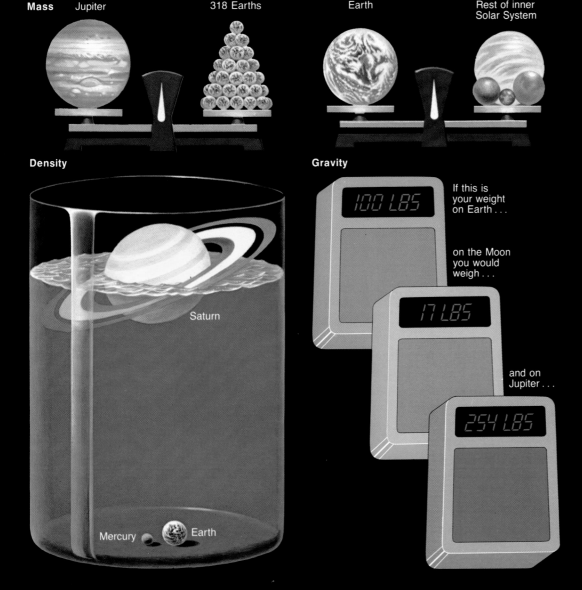

Mass Jupiter — 318 Earths — Earth — Rest of inner Solar System

Density

Saturn

Mercury Earth

Gravity

100 LBS — If this is your weight on Earth . . .

on the Moon you would weigh . . .

17 LBS

and on Jupiter . . .

254 LBS

Mass, density, gravity—what do they mean? *Mass* is the amount of material that something contains. All objects have mass—even you. Mass floating in space weighs nothing, but it acquires weight in the field of gravity belonging to another mass—like a planet. Jupiter's mass equals 318 Earths, and Earth's equals the Moon, Mercury, Venus, and Mars put together.

Density tells us how tightly mass is packed. A dense object feels heavy for its size. A one-pound rock and a one-pound pillow have the same mass, but the rock is denser; it takes up less space. The density of the inner planets averages partway between stone and iron. In our giant water glass, they would sink. But Saturn, mostly gas and ice, is less dense than water; we can imagine it floating.

Gravity: The more mass a planet has, the stronger the pull of its gravity, and the more you weigh there—although the mass of your body stays the same. Your Earth weight falls to one-sixth on our Moon, but more than doubles on massive Jupiter. Yet if Jupiter's mass is 318 times Earth's, why aren't you 318 times heavier at its surface? The reason: gravitational attraction weakens with distance. Jupiter is not dense and compact like Earth, so its surface lies far from its massive center.

	☿ Mercury	♀ Venus	⊕ Earth	♂ Mars	♃ Jupiter	♄ Saturn	♅ Uranus	♆ Neptune	♇ Pluto
Average distance from the Sun (Millions of kilometers)	57.9	108.2	149.6	227.9	778.3	1,429	2,875	4,504	5,900
Revolution	88 Days	224.7 Days	365 Days	687 Days	11.86 Years	29.46 Years	84 Years	165 Years	248 Years
Rotation	59 Days	243 Days Retrograde	23 Hours 56 minutes	24 Hours 37 minutes	9 Hours 55 minutes	10 Hours 39 minutes	17 Hours 18 minutes Retrograde	16 Hours 6 minutes	6 Days 9 hours 18 minutes Retrograde
Average orbital speed (Kilometers per second)	48	35	30	24	13	9.6	6.8	5.4	4.7
Equatorial diameter (Kilometers)	4,880	12,100	12,756	6,794	142,984	120,536	51,100	49,500	2,300
Mass (Earth = 1)	.055	.815	1	.107	317.9	95.2	14.54	17.2	.002
Density (Water = 1)	5.4	5.3	5.5	3.9	1.3	.7	1.2	1.56	.7-.8 (?)
Surface gravity (Earth = 1)	.38	.91	1	.38	2.53	1.07	.91	1.16	.05 (?)
Known satellites	0	0	1	2	16	18	15	8	1

What if . . .

. . . there really were creatures on other planets? Scientists now suspect that in our Solar System only Earth cradles life. We cannot yet rule out Mars, but so far our spacecraft have not detected anything obviously alive. Some experts still hope to find life elsewhere — perhaps under the ice of Jupiter's moon Europa, or even floating in the atmosphere of a gas-giant planet — but chances are slim. Yet we can still try to imagine how life might adapt to the environments of other worlds. Just for fun, let's go on an imaginary safari to real places faithfully described, and see some creatures that never were.

Titan   On dim, cold Titan, Saturn's giant moon, *stovebellies* might live — perhaps by the icy shores of an ethane lake. To avoid freezing, they keep fires burning inside their bodies. How? Stovebellies eat ice, which forms much of Titan's surface. Their fuel is made of oxygen from the ice and methane from the dense atmosphere. By squirting flame like a rocket, they can make long leaps in Titan's low gravity. Amphibious *fishimanders* like to crawl out of the lake and cuddle by a handy stovebelly for warmth — until their host blasts off, sending its guests flying.

44

Mars Whisper-thin winds hiss along a dry, dusty canyon. Deadly ultraviolet radiation pours from an unshielded Sun. Nighttime cold reaches –80°C. Perfect weather for a fellow like the *Martian waterseeker*. Its parasol tail can lift three meters in Mars' low gravity, shading it from ultraviolet sunburn. The long snout can probe for pockets of ice under dried-up channels. And the giant ears, needed to hear well in the thin air, also serve as blankets: In Mars' frigid nights the waterseeker stays snug by clamping its ears tightly around its whole body.

Europa Flat ice covers the second of Jupiter's four major satellites. Europa may be the smoothest globe in the Solar System. And here *brinker-roos* might frolic, on feet shaped like skates. They lead a carefree life, living on pure energy as they zoom across the endless frozen plains. Since there's no air to breathe and no food to eat, brinker-roos need no mouths or noses. Their green skins can carry out photosynthesis in sunlight, as plants do. And the coils on their backs pick up energy from Jupiter's strong magnetic field, which Europa must travel through as it orbits the giant planet.

Pluto   Electrical, crystal beings like these *Plutonian zistles* would find –250°C too hot for comfort. At night, when it's colder still and electricity flows perfectly, zistles feel best. Highly intelligent, they spend most of their time radioing great thoughts to each other. When zistles do get going, they can spring 20 meters high in Pluto's feeble gravity. Zistles think Pluto is the only planet with life — it's too hot everywhere else!

Venus   To survive Venus's heat — lead would melt here — you might need a body that feeds on rock and metal. This *oucher-poucher* (above) snacks on a space probe from Earth. Venus's surface is so hot that oucher-pouchers keep shifting from one foot to the other. They travel by inflating their pouchlike bodies and bouncing along the ground. Every time one lands, it utters its customary cry, which sounds remarkably like "ouch!"

Jupiter   From birth to death, any life in Jupiter's wild atmosphere would have to stay airborne (right) — there's no place to stand. Hanging from their gasbags, floating *jellyblimps* would be easy prey for hungry *swordtails*. A swordtail uses Jupiter's strong gravity and its own pointed body to dive right through its victim. All creatures here must avoid winds blowing toward the freezing layer above or the scorching pressure below.

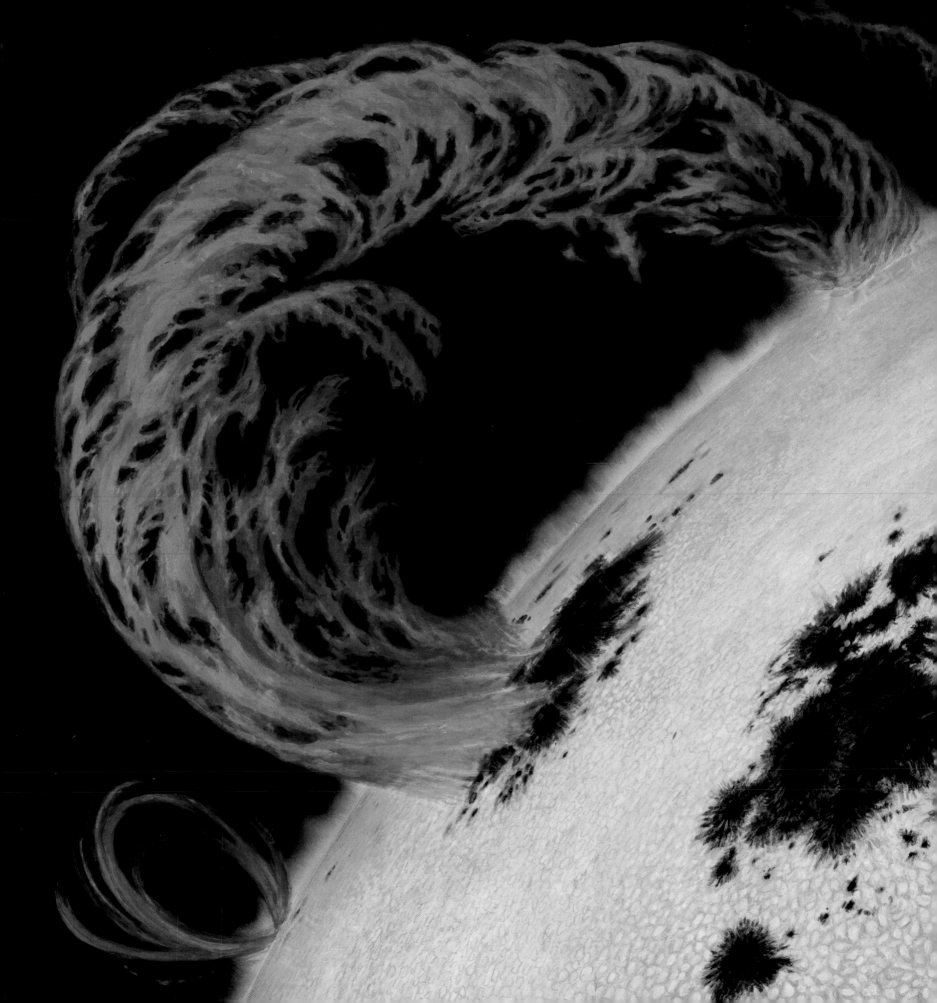

The Star We Know Best

The Sun

The Sun, like a sea of flame, is never still. Restless and seething, it is the star nearest Earth, so it is the star we know best. Here, a great, looping prominence explodes out of a strong magnetic area around a sunspot, hurtles hundreds of thousands of kilometers skyward, then arcs back down. An especially active Sun produces many sunspots, some large enough to gobble up several Earths. Across the surface, hot gases well up in cells called granules. The dancing gas spikes that seem only to rim the Sun actually cover its entire surface and form the lower atmosphere. The outer atmosphere, the corona, extends millions of kilometers beyond the surface gases. A tuft of the prominence has been caught in the Sun's magnetic field.

In ancient times, people thought of the Sun as a perfect sphere of celestial fire created by the gods. Later it was seen as a solid object or a ball of liquid. Sunspots were a puzzle to astronomers of old: Were they clouds? Were they mountain peaks? Were they windows to a cool surface? We will answer these questions as we descend through the Sun's atmosphere, through its gaseous "surface," and into the core where its great nuclear furnace generates the energy that makes possible our life on Earth.

Facts about
The Sun

Apollo the Sun god brings life-giving heat and light to Earth. As patron god of musicians and poets, he carries a lyre. Symbol: ☉ the egg of creation.

Birth and death of the Sun: About five billion years ago, the Sun began to form in a huge cloud of gas (right). As the material condensed, high temperatures and great pressures built up at the center. This set off a nuclear reaction that still releases energy and causes this star to shine. In another five billion years, as the Sun's hydrogen is used up, it will expand to the *red giant* stage, swallowing Venus and Mercury and making Earth's surface semi-molten. The outer layers will expand into space, leaving a *white dwarf*. When the Sun cools, only a *black dwarf* cinder will remain.

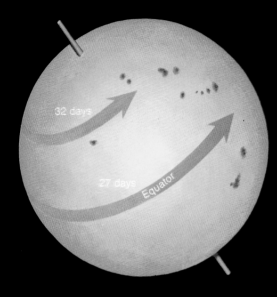

Gamma rays

X rays

Ultraviolet

Visible light

Infrared

Microwaves

Radio waves

Tilted about 7° to the plane of Earth's orbit, the Sun spins on its axis from left to right, as we see it from Earth. By tracking dark specks called sunspots, astronomers first observed that the Sun rotates unevenly. Sunspots near the equator are carried faster than sunspots near the poles. This uneven rotation produces strong magnetic fields and leads to violent solar storms. The Sun's surface bubbles and explodes constantly in a raging inferno of gases.

Radiation from the Sun travels in waves (right). Each wavelength—from gamma rays to radio waves—carries a different amount of energy. Only light, the wavelengths that make up the band of rainbow colors, is visible to the human eye. But we can feel infrared rays as heat, and ultraviolet rays tan our skin. Some radio waves reach Earth, heard with special instruments as a low static noise. The other radiation does not pass through Earth's atmosphere.

WARNING!

Never look at the Sun directly or through a telescope or binoculars. If you do, you can damage your eyes permanently.

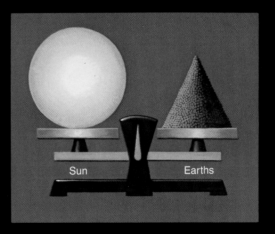

Sun   Earths

Mass of the Sun: It would require about 333,000 Earths to equal the Sun's mass. But more than a million Earths could fit inside. This is because the Sun's volume is greater. Solar matter averages one-fourth as dense as earthly matter.

Comparing sizes: The Sun is an average size star, but it dwarfs our planet (below). About 109 Earths could fit side by side across the diameter of the Sun.

Sun
Diameter: 1,392,000 km

Earth
Diameter: 12,756 km

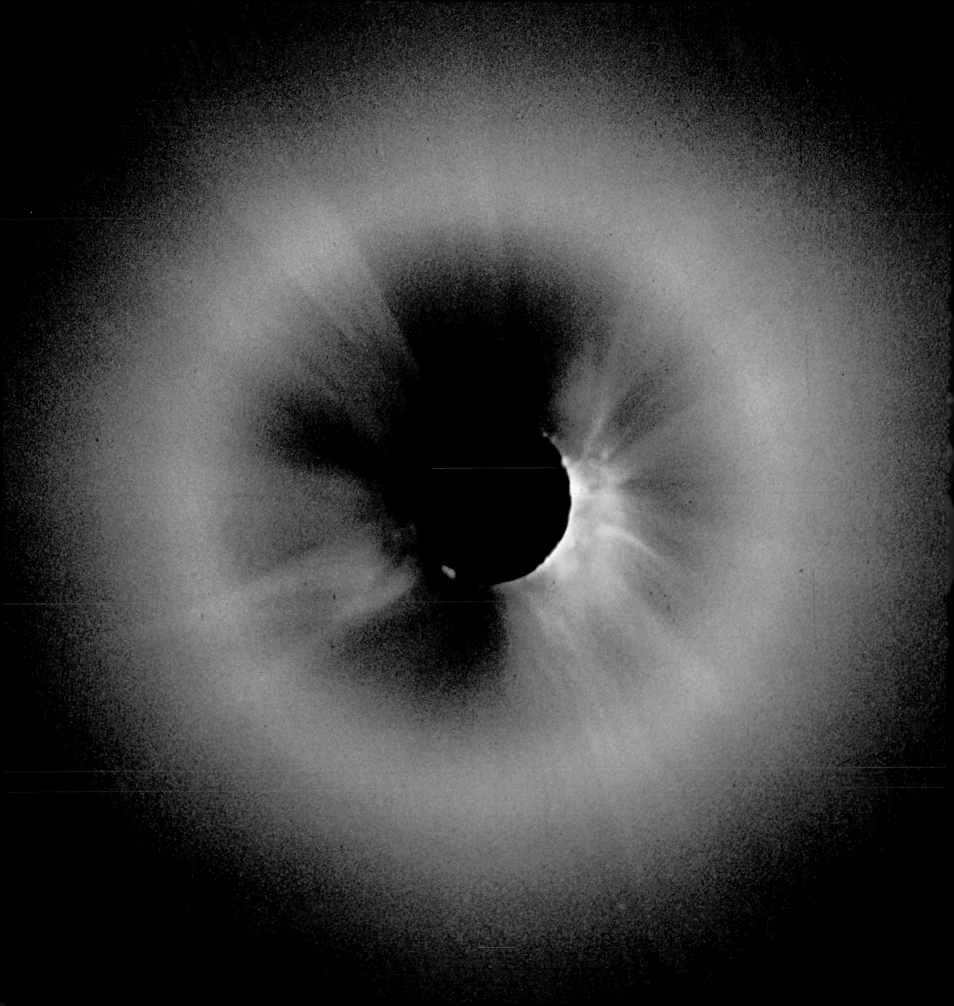

The Sun is always changing. And its changes are what interest us most as we try to learn about our local star—what it is made of, what makes it shine, and how long it will keep shining. The more we find out about the Sun, the more we will know about other stars. Another reason to study the Sun is that it shapes our lives on Earth in many ways. Weather and climate are only two of them. So the more we know about the Sun, the better we can predict its changes and their effects on Earth.

By bouncing radar signals off the Sun and timing their return, we can measure the Sun's distance from us—about 150 million kilometers. If we could fly to the Sun in a jet airliner traveling 1,000 kilometers an hour, the journey would take us about 17 years. But if we traveled at the speed of light—300,000 kilometers a second—we would make the trip in only eight minutes!

A journey through the atmosphere

During a total eclipse of the Sun, the Moon blocks the glaring disk. Then we can see the Sun's atmosphere fanning out into space. The main outer part of this atmosphere, the *corona*, glows in a frail, flame-like pattern. Its brightness is about that of the full Moon.

The corona begins about 2,500 kilometers above the surface and stretches out beyond the orbit of Earth. But the corona's shape changes from day to day. Sometimes it swells up and shoots hot streamers of matter across millions of kilometers.

Photographs of the corona usually show fanlike patterns of gas at opposite poles of the Sun. These patterns are evidence that

the Sun has a magnetic field. This great magnetic field is made of many small magnetic areas which force their way out and spread over the surface. These slowly changing fields act together to vary the corona's shape over days or weeks.

The Sun is composed mostly of hydrogen, with some helium and other elements. There are no whole atoms inside the Sun, since high temperatures would cause them to collide and smash to bits. Individual pieces swim about freely as a thin solar soup called *plasma*. Tons of this plasma are cast off by the Sun each day and stream into space as the *solar wind*.

Gas storms in the chromosphere

Sandwiched between the corona and the Sun's surface is the lower atmosphere, the *chromosphere*, or "color sphere." During an eclipse we see it as a thin, ragged, pinkish rim of light. This is a denser layer of gas than the corona but it, too, is almost empty of matter—a near vacuum. The temperature of the chromosphere is much lower than that of the corona because its atomic particles are moving about more slowly.

The chromosphere is torn by eruptions. Its entire surface dances with bursts of gas 1,000 times higher than the highest mountains on Earth. Called *spicules*—"little

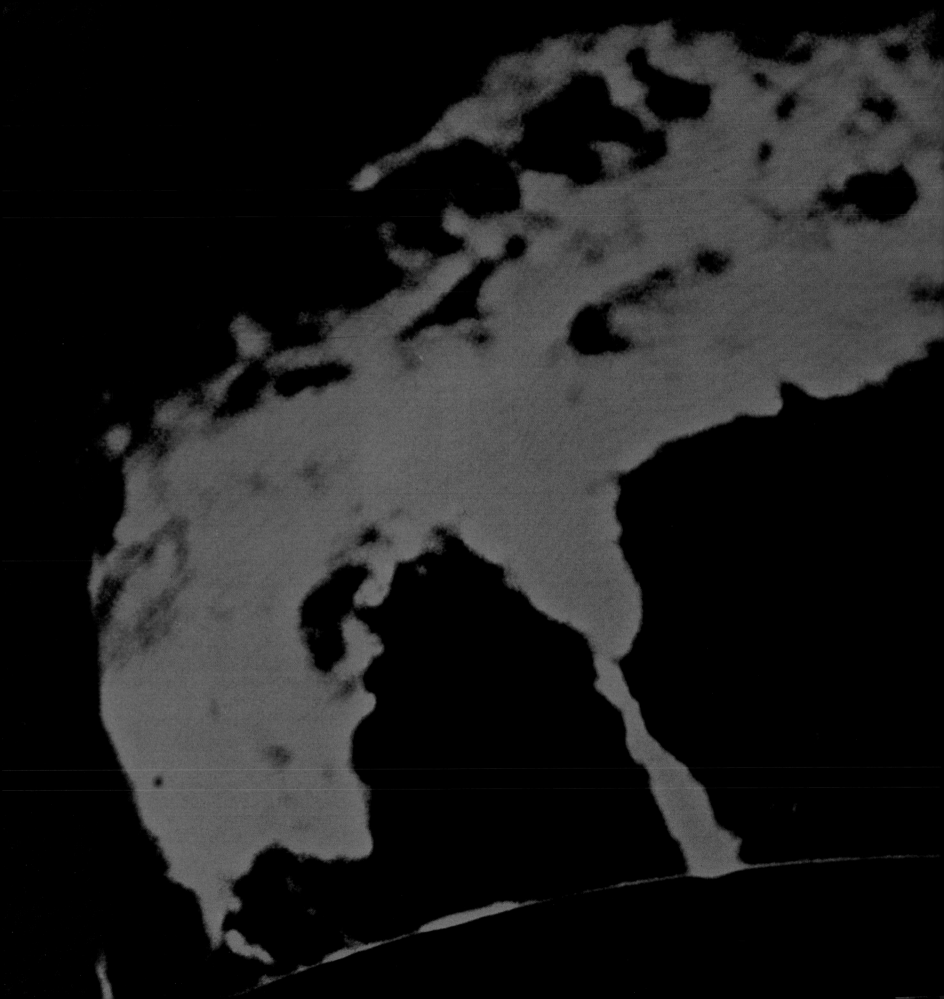

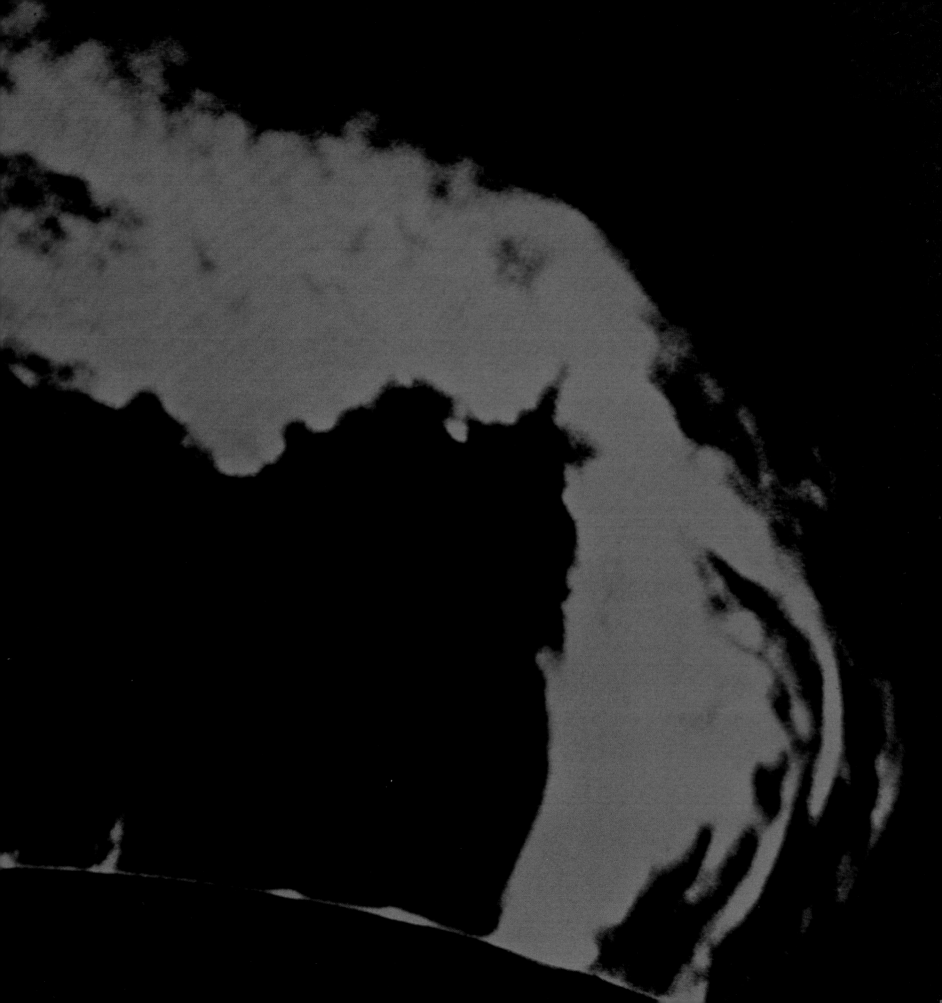

Close-up: A powerful solar telescope zeros in on sunspots below the Sun's seething chromosphere. In this hydrogen-alpha photograph, they appear dark because they are cooler than surrounding gases. At lower left, a violent but short-lived eruption called a flare surges upward.

Sunspots vary in number on an 11-year cycle (bottom). In 1980 NASA's Solar Maximum Mission found that these powerful magnetic fields block some of the Sun's energy, perhaps affecting weather on Earth. But what happens to the energy? Scientists debate this puzzle.

spikes"—these pointed eruptions encircle large cells of upwelling gas. Fenced in by spicules, each cell of gas measures about 30,000 kilometers across and is called a *supergranule.* A supergranule may last half a day. It wells up, spreads out, and dissolves. Then a new one replaces it.

The most violent events on the Sun are *flares,* fiery eruptions like the flash explosion of a huge pool of gasoline. Just one sets free the energy of 10 million hydrogen bombs. When a flare erupts in the chromosphere, the energy of twisted magnetic fields is suddenly released as intense light and streams of particles. When the solar surface is especially active, astronomers see flares every hour or two. But when the Sun is quiet there may be none for days or weeks. Flares help to cause the "northern lights," or *aurora borealis,* and can disrupt shortwave radio signals.

Great clouds of gas—*prominences*—hang above the chromosphere, held by magnetic fields. As they come out through the surface, the magnetic fields continually stretch and change, sometimes hurling vast amounts of plasma high into the solar sky. Gigantic geysers of glowing gas may surge upward more than 100,000 kilometers at speeds as great as 500 kilometers a second. These prominences often make graceful loops. Sometimes they float for several months, like summer clouds—but clouds at temperatures as high as 10,000 kelvins. (Kelvins are units of temperature; room temperature is about 300 kelvins.) Some eventually cool and fall back into the Sun. Others are thrown out so fast that they escape from the Sun's

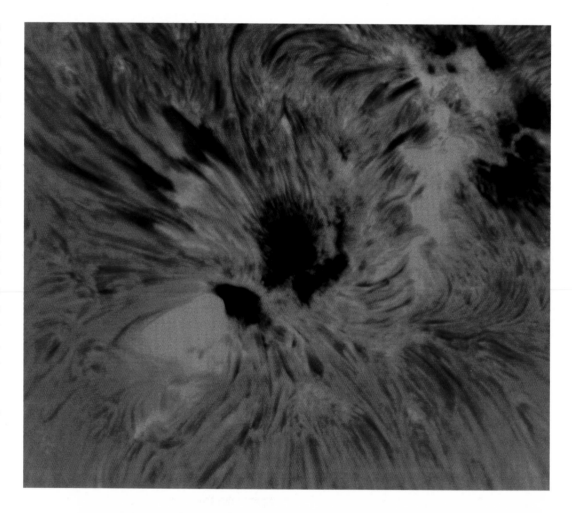

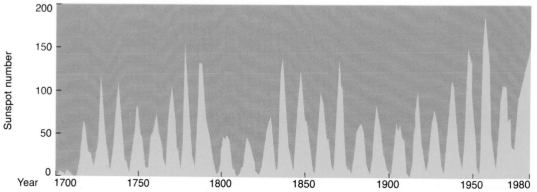

How sunspots form: According to the "twisted rope" concept, a pair of sunspots may start with a single line of magnetic force lying north-south. (1) Because the Sun rotates faster at the equator, the middle of the line progresses faster. (2) The line has wrapped several times around the Sun. (3) The line has stretched and twisted like a rubber band, creating very strong local magnetic fields. Finally, a kink of tangled magnetic field erupts through the surface and produces a pair of sunspots. Flares and prominences form in the sunspot area.

1

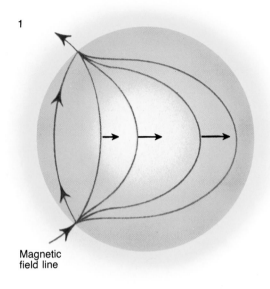

Magnetic
field line

2

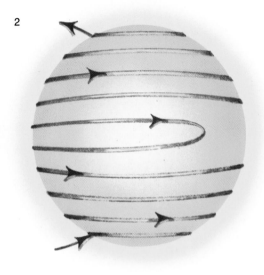

3

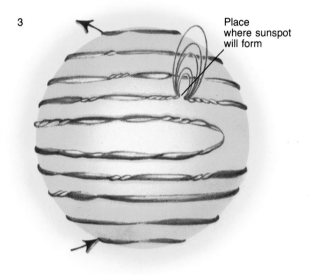

Place
where sunspot
will form

gravitational pull and are lost in space.

Beneath the chromosphere is the Sun's surface, called the *photosphere,* or "sphere of light." It is a bubbling sea of hot gas cells called *granules.* These are like the chromosphere's supergranules but much smaller and with a life of only about six minutes. The photosphere gases have a temperature of about 6,000 kelvins. Most of the light and other energy we receive comes from the Sun's thin photosphere layer.

Sunlight

The sunlight we see every day is made of tiny units of radiant energy called *photons.* Born in the inferno of the Sun's core, they spend millions of years slowly wandering up to the surface. Then in eight minutes they speed across the 150 million kilometers to Earth—if they happen to be headed in our direction. Depending on the amount of energy a photon has, it may be absorbed by Earth's atmosphere. Or it may zip down

to Earth's surface and help warm a flea or a blade of grass for a fraction of a second. Each photon carries only a tiny amount of energy but trillions of them hit each square meter of Earth every second. Together they form sunlight. Of course, clouds, water, and the ground itself reflect a lot of sunlight back into space. It is this reflected sunlight that allows us to see Earth from a spaceship.

Spots on the Sun

More than 2,000 years ago, Chinese astronomers reported seeing sunspots when thin clouds, smoke, or dust dimmed the Sun's blinding disk. In 1612 the Italian astronomer Galileo studied them with the help of a telescope. By focusing the Sun's image on a piece of paper, he could trace the spots. (No one should ever look directly at the Sun, especially with a telescope.) Day after day he watched them glide across the Sun's face near the equator. Each spot took

about 13 days to move from one edge and slip around the opposite edge. This movement told Galileo that the Sun rotated.

Since Galileo's time, people have kept track of sunspots. When there are a lot, we say that the Sun is active, and when there are few, we say it is quiet. An active sunspot period begins when several spots break out about midway between the poles and the equator in both hemispheres of the Sun. Then year after year the new spots emerge closer and closer to the equator.

A large sunspot may be five times larger than Earth. Around the edge, thin gas streamers make a pattern of magnetic force lines. This suggests that each sunspot is the location of extremely powerful magnetic fields—several thousand times stronger than Earth's average magnetic field. Such powerful magnetic fields prevent the usual flow of energy upward, making the Sun's surface cooler and thus darker in the area we call a sunspot. Thus sunspots are not the

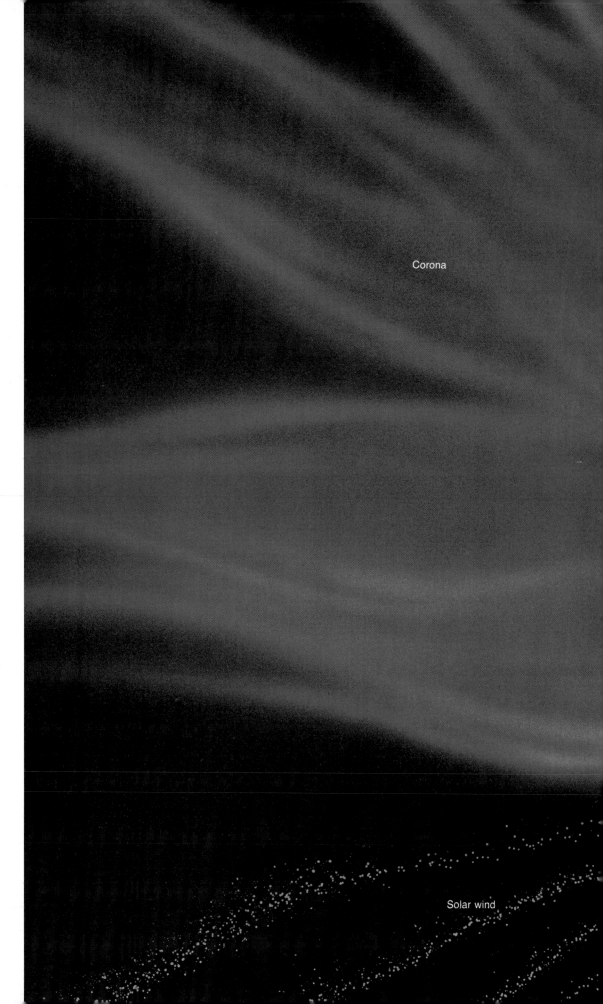

Portrait of the Sun shows its most widely studied features and a cross section of its interior. From the core, nuclear energy radiates out to the convective zone, where giant bubbles of hot gas rise and sink. The chromosphere and corona make up the Sun's lower and outer atmospheres.

raging tornadoes people once thought. They are giant magnetic fields that cool areas of the Sun. Also, sunspots contain the strong magnetic forces that ignite solar flares, trigger prominences, and change the shape and size of the corona. When there are many sunspots, there are many flares and prominences. Sunspots announce the coming of storms on the Sun. They are a storm alarm for the entire Solar System.

The sunspot cycle and climate

The Sun's activity seems to bring about changes in Earth's climate. From about 1645 to 1715 few sunspots erupted—the Sun was fairly quiet—and those 70 years were unusually cold. They were part of a period later called the Little Ice Age, which extended from about 1400 to 1850.

Astronomer John A. Eddy has shown that for 7,000 years glaciers on Earth have advanced and retreated in step with the Sun's activity. When the Sun is very active, the glaciers retreat. When it is fairly quiet, they advance again. So sunspots may well have far-reaching effects on all of Earth's life, effects we are only now beginning to learn about.

How does the Sun shine?

If we could make our way deep beneath the boiling photosphere, we would find three things happening. One: The temperature steadily rises, to some 15 million kelvins in the core. Two: The weight of the bits and pieces of atoms pushing from above creates extremely high pressure. Three: Core matter packs so tightly that it is about 10 times denser than silver or iron.

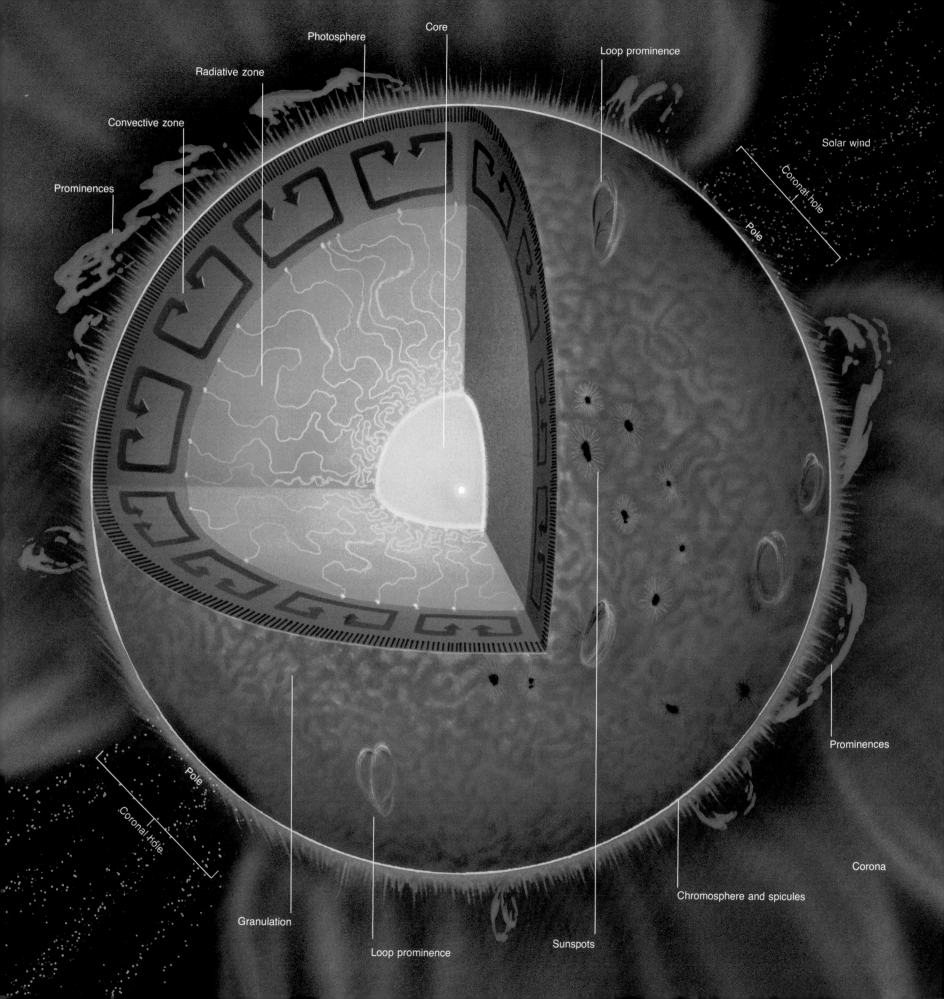

Convective zone

Radiative zone

Photosphere

Core

Loop prominence

Prominences

Solar wind

Coronal hole

Pole

Prominences

Corona

Pole

Coronal hole

Granulation

Loop prominence

Sunspots

Chromosphere and spicules

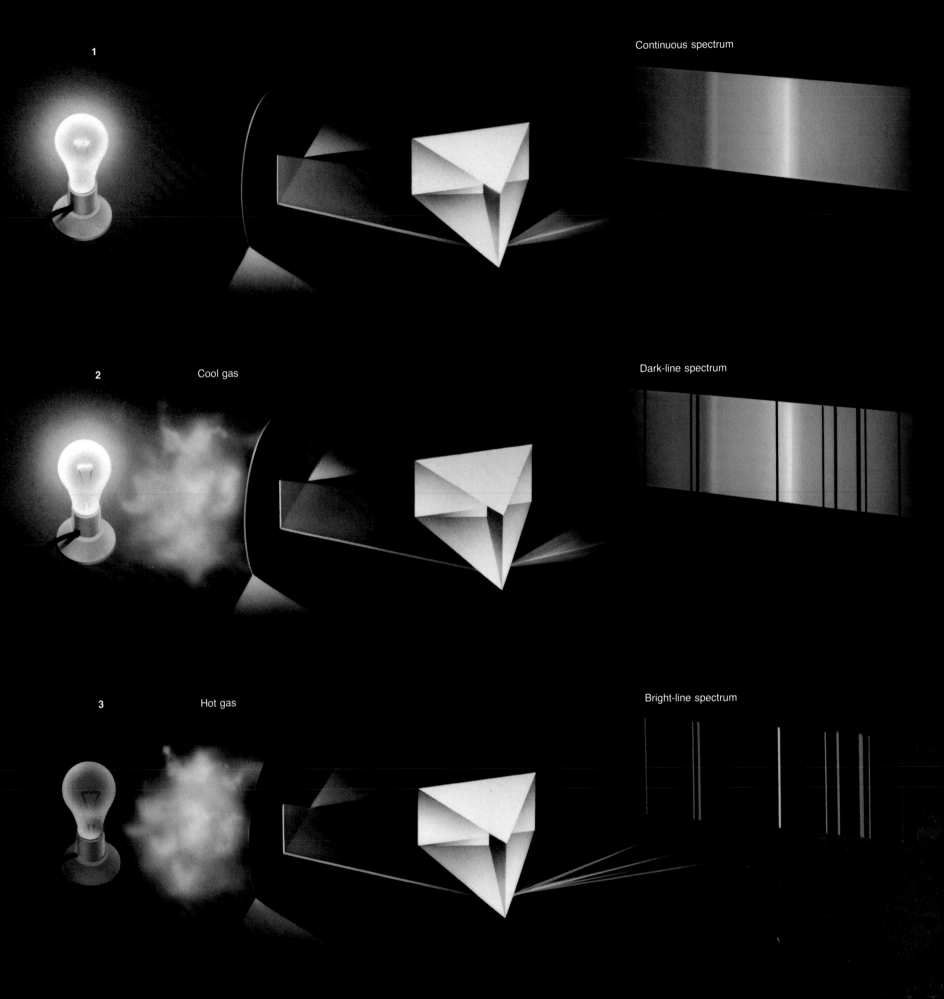

1

Continuous spectrum

2 Cool gas

Dark-line spectrum

3 Hot gas

Bright-line spectrum

Scientists decode starlight with spectra. Light from a bulb's hot filament passing through a prism produces a continuous band of colors (left). If the light goes through a gas, certain wavelengths are absorbed, resulting in dark lines. A hot, glowing gas creates bright lines.

Sunlight produces a dark-line spectrum (below) because it passes through gases in the Sun's atmosphere. Each line identifies a particular form of chemical element. For example, the H-alpha line of hydrogen appears at the far right on the top red band. A strongly marked pair in

the middle of the second red band indicates the presence of sodium. Magnesium is identified by three lines in the green region of the spectrum. These dark lines in the solar spectrum are called Fraunhofer lines in honor of the man who first studied them carefully.

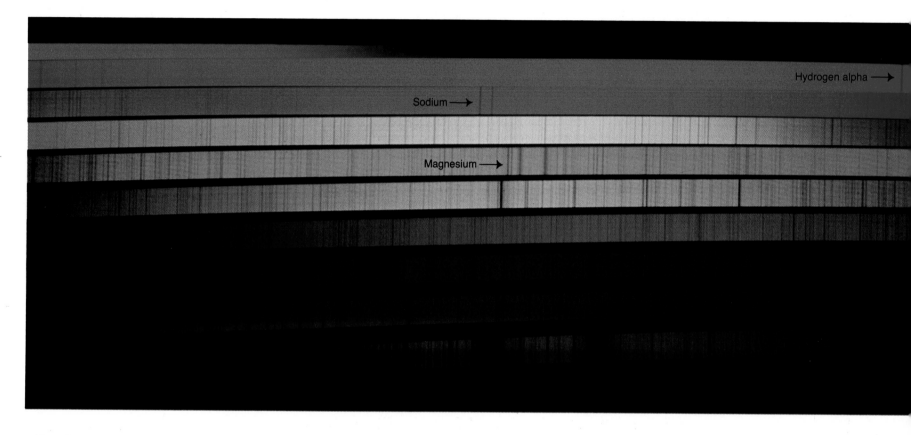

Hundreds of years ago people thought the Sun was a ball of fire. Today we know it cannot be. A lump of coal the size of the Sun would burn out in a few thousand years, and the Sun is much older than that. Another theory was that the Sun gave off its tremendous energy by shrinking and packing its matter tightly around the core; a star in its forming stages does that. But the Sun does not shine that way now. It would have shrunk to nothing long ago.

Do we have an answer? In the early 1900's Albert Einstein said that matter—a block of wood, a piece of chalk, you, anything—can be changed completely into en-

ergy. This was a startling idea, but one that we now know is true. Another scientist, Hans Bethe, used this idea in the 1930's to suggest how the Sun has been producing energy. Bethe received the Nobel Prize for this work in 1967. To understand Bethe's theory and others since, we must revisit the Sun's inner core—a huge ball twice the size of Jupiter. Here the great pressure and the high density and temperature cause the crowded jumble of atomic particles to smash into each other violently and often. Sometimes certain of these particles ram into each other so hard that they stick together in a process called *fusion*.

Our H-bomb Sun

Here is what we think happens: Two hydrogen *nuclei*, or *protons*, smash into each other hard enough to fuse, forming a deuterium nucleus with one proton and one *neutron*. As the protons fuse, they release energy, partly in the form of a *neutrino*. Neutrinos are odd particles that seem able to penetrate almost anything, including us and Earth as if neither existed.

Next the deuterium nucleus rams another free proton and fuses with it, giving off energy in the form of gamma rays. Finally, this clump of three particles smashes into another clump like itself and forms

Giant Sun-watcher, the McMath Solar Telescope at Kitt Peak, Arizona, stands 11 stories high and slopes 50 stories down into the ground. Atop the tower, movable mirrors follow the Sun, reflecting its rays down the slanting tunnel. Other mirrors focus the light to form the Sun's image.

a helium nucleus. In this last collision two protons are knocked loose and the process begins again. In each step, energy is given off in the form of photons. It is this energy that keeps the Sun shining and Earth alive. And this is how a hydrogen bomb works. The Sun is an enormous hydrogen bomb that just keeps on exploding.

A 50-million-year trip

How does the energy get out of the core? Neutrinos whiz up at the speed of light, but most of the energy takes a long, winding route to the surface. Scientists call this a "random walk." By the time a bundle of the original core energy reaches the photosphere, 50 million years have passed. That bundle, which started as gamma rays, now contains photon energy all along the electromagnetic spectrum. (See pages 28-29).

In one second the Sun gives off more energy than all people have produced during their entire stay on Earth. It is hard to believe that a tiny amount of mass can produce such a large amount of energy, but it can. If a piece of any matter weighing only one kilogram could be turned completely into energy, it would supply all of the electricity needed by the entire United States for two months! Yet our planet receives only about two-billionths of the total energy output of the Sun. The rest streams out in all directions into space. On the average, each square meter of Earth's surface receives enough solar energy to heat and light one small room. If only we knew how to capture more of that energy.

If the Sun is using up its hydrogen to produce energy, won't there come a time when it runs out of hydrogen and goes out? Such a time must one day come. But even though the Sun uses four million tons of its hydrogen fuel every second, it has enough to keep shining for five billion more years.

The message of sunlight

When fitted to a telescope, a *spectroscope* acts like a prism. It spreads light out in the rainbow band called a *spectrum*. Once we can read that spectrum, it reveals dozens of secrets about the Sun or any star.

If light from a glowing bulb passes through a slit and then a prism (see page 60), we see the smoothly graded band of colors called a *continuous spectrum*. Now, if we put a gas between the bulb and the slit, a pattern of dark lines crosses the spectrum, which is no longer "continuous." This is a *dark-line* or *absorption spectrum* (because the gas has absorbed some specific wavelengths of the light energy). Every chemical element produces a unique "fingerprint" pattern of dark lines.

Now, let's turn off the light bulb and look at the spectrum of the same gas after it has been heated. Where there was a pattern of dark lines before, there is now a pattern of brightly glowing lines called a *bright-line* or *emission spectrum*. The gas is *emitting* — putting out — its fingerprint.

When we look at the Sun through a spectroscope, it is like looking at light from the bulb shining through a gas. The gases of the Sun's atmosphere lie between the Sun's surface and our spectroscope. So we see a dark-line spectrum and can identify the fingerprint patterns of hydrogen, helium, iron, magnesium, and about 70 other

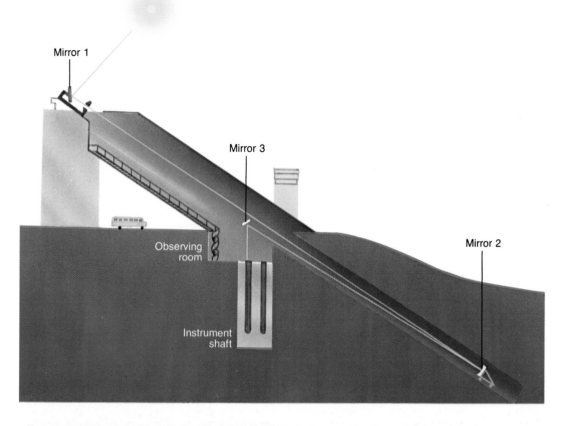

Cutaway drawing at left shows how the solar telescope works: Sunlight striking Mirror 1 travels down the diagonal shaft. A second, image-forming, mirror at the end sends the light back up to a mirror at ground level. This third mirror projects the beam to the observing room and to the instrument shaft. On top of the tower (far left), astronomer Bruce Gillespie adjusts the position of the large tracking mirror, called the heliostat. Inside the telescope (below), Bruce watches Mirror 3 as he lines it up with the concave mirror at the far end of the tunnel. Dark glasses shield his eyes from the Sun's glare. (See page 66 for more on solar telescopes.)

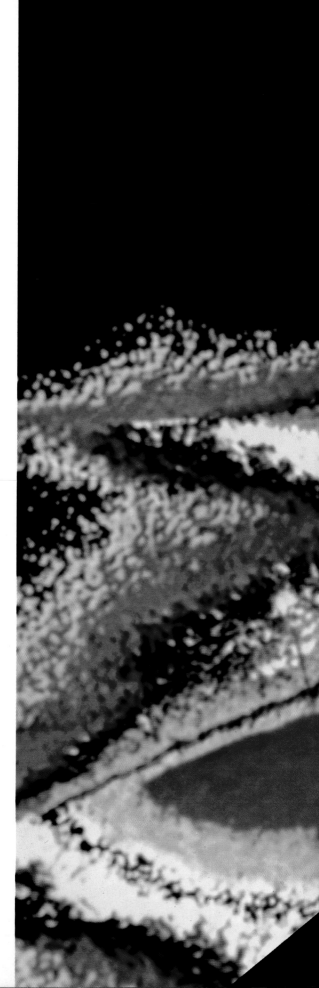

elements of the chromosphere and corona.

Spectra show much more than what the stars are made of. They tell us the temperature of the stars' gases, their pressure, density, and motion. They even show us where magnetic fields are located.

Solar telescopes

Astronomers use telescopes with very long focal lengths to study the Sun. The McMath Solar Telescope at Kitt Peak, Arizona, the largest in the world, gives an image of the Sun that is almost one meter across. It shows features such as sunspots in great detail. Because it is the only telescope of this magnitude ever built, the McMath draws astronomers from all over the world. Stable air temperature is important in using a telescope, so the McMath has an inner skin of 36 tons of solid copper tubing. Some 64,000 liters of antifreeze flow through this tubing and prevent hot air currents that would blur the image.

Behind the McMath telescope (pages 62-65), is a smaller vertical tower containing a solar vacuum-telescope. Its long vacuum shaft has practically no air in it, producing a sharper, though smaller image. This telescope makes daily magnetic maps—or *magnetograms*—of the Sun.

The McMath operates 24 hours a day, observing not only the Sun but the Moon, planets, and stars. Instruments such as the McMath do double duty by enabling scientists to study the makeup of both sunlight and starlight. Another Kitt Peak telescope, the Mayall, produced the picture of the star Betelgeuse on page 226.

The many faces of the Sun

Viewing the Sun through a spectroscope is something like viewing a landscape through binoculars. With binoculars we can first focus on nearby objects. By changing the focus we can bring into view objects farther away. At any moment we see clearly only those parts of the landscape we choose to keep in sharp focus. The others are either fuzzy or invisible.

By "focusing" the spectroscope at red wavelengths in the visible spectrum, and blocking out other wavelengths, we photograph cooler, low-energy gases in the lower photosphere and nothing else. When we slowly change the focus up through the shorter and shorter wavelengths of orange, yellow, green, blue, and violet, to ultraviolet and X rays, we can examine the Sun's atmosphere layer by layer, photographing each as we go. By the time we reach the ultraviolet and X rays we are focused on the hot corona above the chromosphere.

NASA's first experimental space station, Skylab, was fitted with ultraviolet and X-ray telescope-cameras that gave us many thousands of pictures of the Sun— views never before possible. Skylab's ultraviolet telescopes photographed the very energetic gases of the high upper chromosphere. The X-ray telescopes photographed the even more energetic gases with temperatures of more than one million kelvins in the high corona. For the first time we saw the Sun's coronal holes. The coronagraph, ultraviolet, and X-ray segments in the composite photograph shown here were taken by telescopes carried on Skylab.

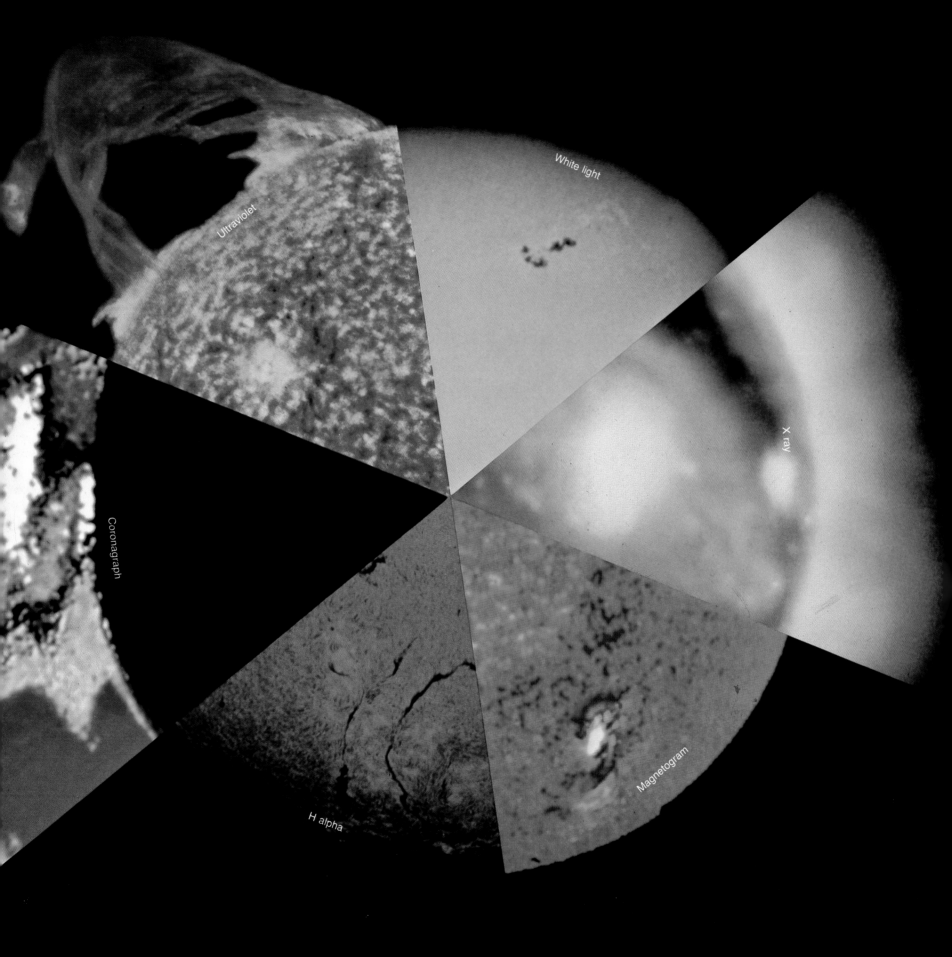

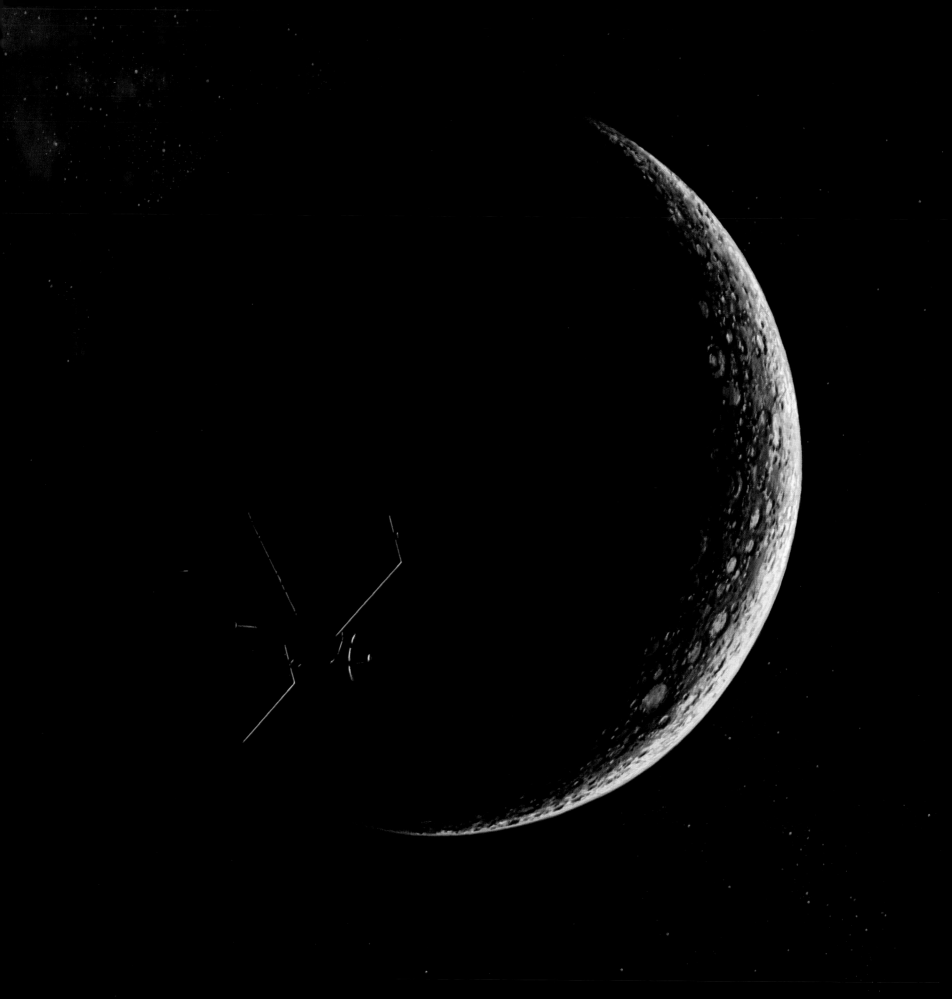

Mercury

The spacecraft Mariner 10 silently glides past the night side of the innermost planet in our Solar System. Mariner's television eye examines the battered surface of Mercury from thousands of kilometers away. It sent back our first detailed pictures of Mercury's Moon-like surface in early 1974.

One of the spacecraft's most important findings was that Mercury has a magnetic field strong enough to turn aside the mighty solar wind. Another was that Mercury's surface is covered with craters. These probably were made early in Mercury's history, when rocky meteorites up to several kilometers across crashed into the young planet and produced the landscape that Mariner's cameras photographed. Between many of the craters are smooth plains, perhaps volcanic in origin, and huge slopes—*scarps*—that formed in the ages after the bombardment ended.

If we could face the Sun from Mercury, the way Mariner's solar energy panels do, we'd see it covering an area in the sky nine times larger than it does from Earth. We would see tufts of solar gases caught up in the Sun's strong magnetic field. We would also understand why a summer vacation on Mercury would not be much fun.

Facts about
Mercury

Snakes entwined on his staff protect Mercury, messenger of the Roman gods. With imagination, you can see the snakes entwined in Mercury's symbol, too: ☿

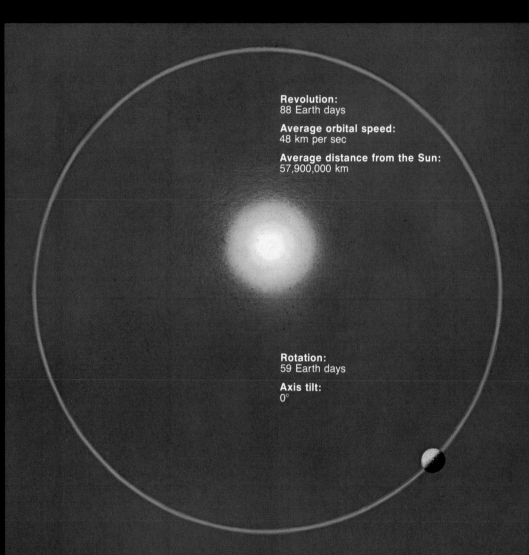

Revolution:
88 Earth days

Average orbital speed:
48 km per sec

Average distance from the Sun:
57,900,000 km

Rotation:
59 Earth days

Axis tilt:
0°

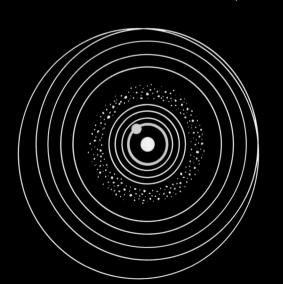

Innermost of the planets, Mercury swoops to within 46,000,000 km of the Sun, then slows down as it swings out to some 70,000,000 km away. Mercury's orbit is more elliptical than the path of any other planet except Pluto. And if orbits were racetracks, speedy Mercury would leave the other planets behind. It zips along at about one and a half times the speed of Earth. But, for all its speed, its spin is slow — so slow that six Earth months go by before the Sun moves from one high noon on Mercury to the next high noon.

Telescopes see Mercury in phases like our Moon's because the planet's orbit lies inside our own. But they see it poorly, for Mercury stays close to the Sun—lost in its glare by day, blurred by our atmosphere when viewed low in the sky at dusk, gone with the Sun by night.

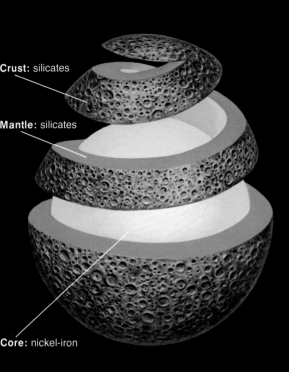

Crust: silicates

Mantle: silicates

Core: nickel-iron

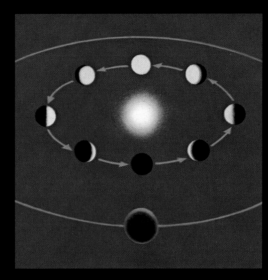

Earth

Mercury
Diameter:
4,880 km

Mercury's rocky rind hides an iron core about as big as the Moon. Twice as iron-rich as any other planet, Mercury is almost as dense as Earth. That makes its gravity about a third that of Earth, though its mass is only one-eighteenth.

A thermometer on Mercury would show freezer-to-furnace extremes in a single day at the equator. At sunrise, an imaginary thermometer hovers at –183°C, almost 300° below zero on the Fahrenheit scale. No spot on Earth gets

nearly so cold. Some 22 Earth days later it's mid-morning on Mercury and about as warm as a summer day here. But in another Earth month, Mercury at early afternoon would melt lead with a peak temperature above 800°F.

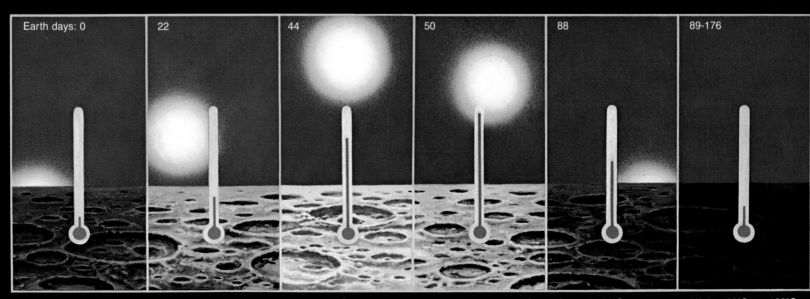

Earth days: 0	22	44	50	88	89-176
Before dawn –183°C	**Mid-morning** 27°C	**Noon** 407°C	**Early afternoon** 427°C	**Sunset** –23°C	**Night** –23°C to –183°C

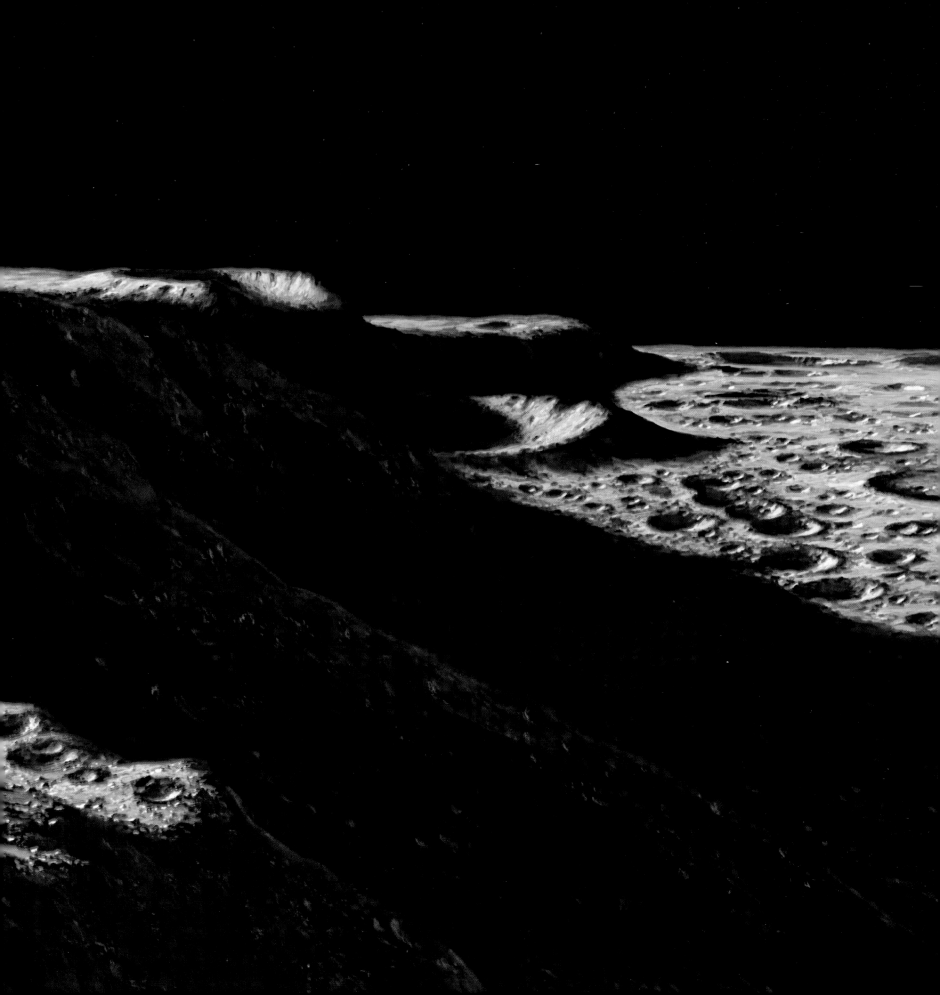

Mercury to within 705 kilometers. Mariner 10 continued on around the Sun, returned for a second flyby, looped around the Sun again, and came back, a year after its first flyby, for a third look at Mercury.

One of the main objectives of the Mariner 10 team was to find out what kinds of surface features Mercury has. In the three visits, Mariner television cameras photographed about half of the planet. These pictures were put together in sections, and maps such as ours were made from them.

The spacecraft also discovered that Mercury has a magnetic field. Like Earth's magnetic field, Mercury's is pushed out of shape by the steady blast of atomic particles from the Sun called the solar wind. The first flyby had detected the magnetic field. The third confirmed that it is squashed up close to Mercury on the sunlit side and stretched out into a long tail on the side away from the Sun.

Although weak—about one percent the strength of Earth's—the magnetic field of Mercury is much stronger than that of Mars. Perhaps it is strong enough to capture a very thin veil of gas given off by the Sun. However, because the gas is of such a low concentration, we cannot call it an "atmosphere." Our own atmosphere at 1,000 kilometers altitude has the same pressure as Mercury's at the surface. That is more than a million times lower than the pressure on Earth's surface. Mercury's veil is far too thin to block out even a little of the Sun's radiation. And, at such a close distance to the solar furnace, Mercury's surface gets many times more radiation than Earth's surface does.

The hot and cold planet

At *perihelion*—the planet's closest approach to the Sun—Mercury is only about 46 million kilometers away from it. This is close enough that, on the equator in early afternoon, the surface temperature reaches 427°C. That's more than twice the temperature it takes to bake a cake. But on the opposite, shaded surface the temperature plunges to –183°C. How can it be so hot and so cold at the same time?

Temperatures rise on Mercury's sunlit side partly because the solar radiation is so intense and partly because Mercury days are so long—from sunrise to sunset each one lasts nearly three Earth months. This is a long time for the crustal rock to heat up. But Mercury's night side does not see the Sun for the same long period of time. So it gets very cold. On Earth, the dense blanket of air shields the planet's day side from some of the radiation, and holds in heat on the night side.

Mercury's high orbital speed and slow rotation (each is a result of the Sun's gravitational attraction) produce the long days and nights. The planet takes 88 Earth days to complete one revolution around the Sun.

Until recently, scientists thought that Mercury also took 88 days to rotate once about its axis, which would make a day on Mercury equal to its year. They thought that the planet always kept its same face toward the Sun. This would have given it a permanent day side and a permanent night side. But in 1965 astronomers at the Arecibo Observatory in Puerto Rico beamed radar impulses at the planet, measured the returning impulses, and determined that

Mercury rotates once about its axis in 59 days. This means that every time Mercury makes two revolutions around the Sun, it has made exactly three rotations on its axis. Because of its short year, you would be one year older about every 88 days if you were a Mercurian. A 15-year-old, by Earth time, would be 62 years old on Mercury, and could expect to live to a ripe old age of 300 Mercurian years.

A Sun that seems to do tricks

From some spots on Mercury we could see the Sun rise in the east and climb for about a month and a half of Earth time. Just before high noon it would seem to "loop"— slow down, stop, back up, stop again, and then resume its westward movement until it set a month and a half later. At other places on the planet, we could watch two sunrises and two sunsets every day. These things happen because of the planet's highly elliptical orbit and because at perihelion Mercury's spin speed is slower than its orbital speed.

Although we do not look to Mercury as a place to live, scientists are especially interested because it is such a close neighbor to the Sun. They would like to test crust samples. If a spacecraft could land instruments, they might tell whether the planet's inner core is liquid or solid. An orbiting spacecraft could take up the photography and mapping where Mariner 10 left off. Someday, instruments on an orbiter or lander will measure changes in the strength of the Sun's gravitation, the solar wind, and other characteristics of the Sun.

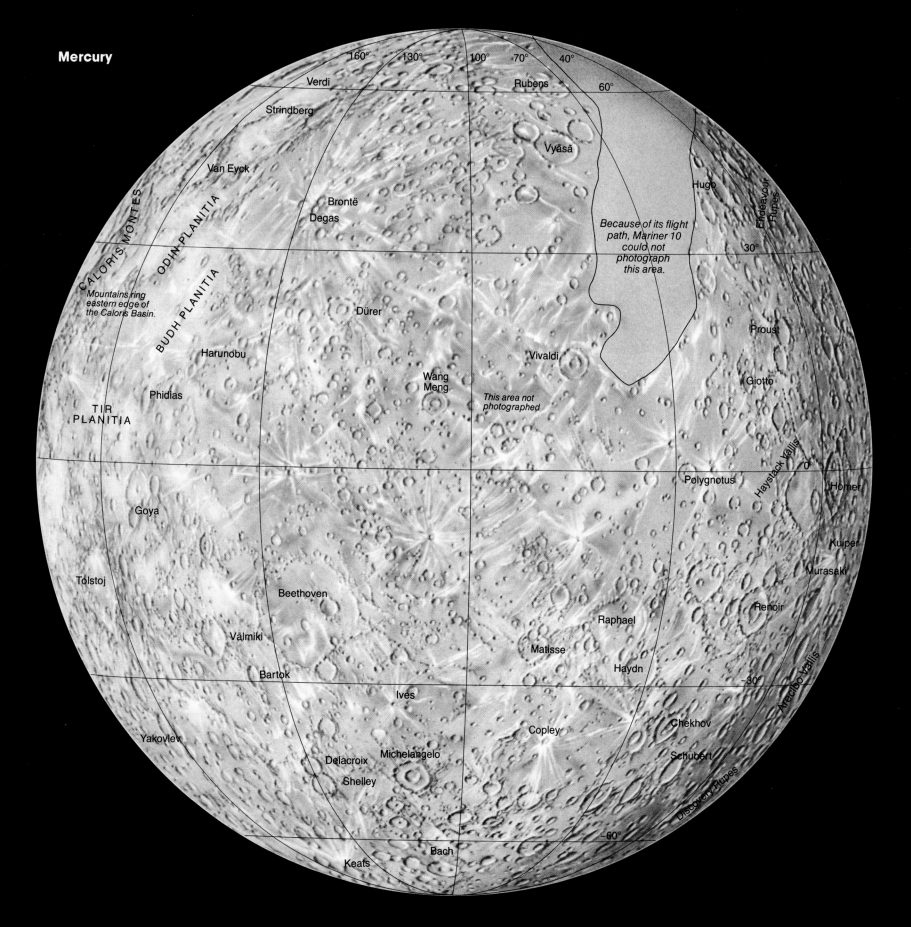

Mercury

160° 130° 100° 70° 40° 60°

Verdi Rubens

Strindberg

Vyāsā

Van Eyck

Hugo

CALORIS MONTES

ODIN PLANITIA

Brontë

Degas

Mountains ring eastern edge of the Caloris Basin.

BUDH PLANITIA

Dürer

30°

Proust

Because of its flight path, Mariner 10 could not photograph this area.

Harunobu

Vivaldi

Giotto

Phidias

Wang Meng

This area not photographed

TIR PLANITIA

Polygnotus

Haystack Vallis

0°

Homer

Goya

Kuiper

Murasaki

Tolstoj

Renoir

Beethoven

Raphael

Vālmiki

Matisse

Haydn

Bartok

−30°

Ives

Arecibo Vallis

Chekhov

Copley

Yakovlev

Schubert

Delacroix Michelangelo

Shelley

Discovery Rupes

−60°

Bach

Keats

Endeavour Rupes

A Veiled Planet
Venus

Venus's upper clouds of poisonous sulfuric acid swirl in a pattern of yellowish mists. Moving at speeds almost three times faster than winds in a hurricane, they race around the planet once every four days. Mariner 10 had this view of mysterious Venus before speeding on to Mercury. Of more than twenty space missions—many American, most Russian—several have given us glimpses of the cloud-veil and the planet's rocky surface.

People once imagined Venus as a Garden of Eden with lush vegetation, sparkling streams, and life. They regarded Venus as Earth's twin because its size, mass, and

density are so much like Earth's. But there the likeness ends. Beneath the acid clouds is a planet we could not imagine in our wildest dreams. Day and night, lightning and thunder flash and boom across a rainless sky. On the surface, the pressure is as strong as the pressure about one kilometer below the surface of the oceans on Earth. The atmosphere is so dense we could nearly swim through it.

Beginning in 1990, the spacecraft Magellan mapped 99 percent of Venus's surface, but some of the planet's mysteries remain: Are there active volcanoes? Were there ever oceans?

Facts about
Venus

Love and beauty, springtime and flowers: the Roman goddess Venus ruled them all. What better symbol for her bright planet than a hand mirror? ♀

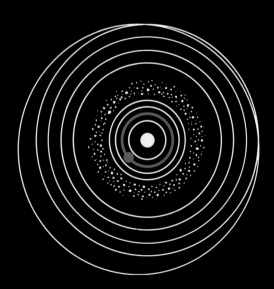

Revolution:
224.7 Earth days

Average orbital speed:
35 km per sec

Average distance from the Sun:
108,200,000 km

Rotation:
243 Earth days
Retrograde

Axis tilt:
3°

Second from the Sun, Venus traces an orbit that is closer to a perfect circle than any other planet's. And yet it moves in odd ways. Its rotation is retrograde, or "backward," so the Sun rises in the west and sets in the east. That spin is so slow that four Earth months go by between one Venus sunrise and the next. And by only the third sunrise, the planet has completed a full swing around the Sun. Thus on Venus a year is less than two days long. But as it orbits, Venus is a speedster at 35 km a second. Only Mercury can outrun that.

The winds of Venus vary with altitude. Upper clouds whip by at 360 km an hour. But a slow breeze at the surface moves no faster than a walk. Venus is a forecaster's dream; its weather scarcely varies. In many places Pioneer and Venera probes logged the wind speeds shown below.

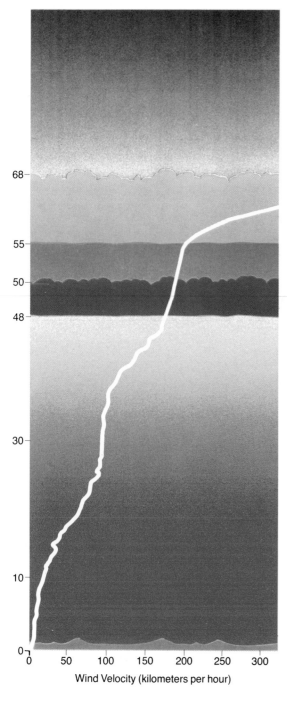

Wind Velocity (kilometers per hour)

and evaporation on Earth. The energy that drives this sulfur cycle may be ultraviolet radiation from the Sun.

On reaching the bottom of this lowest cloud layer we find that the winds have slowed to about 175 kilometers an hour. But the temperature has soared to 90°C, and soon we find the probable source of the radio static that grew louder as we descended—flashes of lightning. Deafening booms punctuate a roar of thunder. In 1990 the Galileo Venus flyby detected signals characteristic of lightning.

Below the clouds we pass through a hazy belt of fine particles. For the remaining 30 kilometers down to the surface the air is very clean. Even so, we have trouble making out details because the atmosphere has become so dense that our view is like one through clear water. A saucer-shaped object drifting down through Venus's atmosphere at this height would zigzag as a dish does when it sinks in water.

Down to the surface

By the time we are 20 kilometers from the surface the temperature zooms to almost 300°C and the atmospheric pressure is a crushing 21 times the pressure on Earth at sea level. It is little wonder that, until now, no spacecraft has been able to survive these conditions very long.

When our imaginary craft finally touches down, it kicks up a cloud of fine dust or sand. Nearly five minutes go by before the dust settles out of the dense air. Other areas, photographed in October 1975 by the Soviet spacecraft Venera 9 and 10 and in March 1982 by Venera 13 and 14, are rocky, with patches of soil. The Veneras returned six pictures to Earth before expiring in the forbidding temperatures and corrosive atmosphere. Those are the only pictures we have of the landscape of Venus from the surface.

Winds on the dry surface are slight, not much more than four kilometers an hour. However, the air is so dense that even this breeze sweeps us along as if it were a river current, and a fiery one. Our temperature sensors now register a blistering 460°C. What makes it so hot?

A carbon dioxide greenhouse

Most of the atmosphere is carbon dioxide, a heavy gas. Nitrogen is present, too, but only 3 percent compared with 78 percent in Earth's atmosphere. But there is little, if any, oxygen. Since carbon dioxide "covers" the planet the way glass covers a greenhouse, we can think of Venus as a planet-wide greenhouse. The atmosphere lets shortwave radiation from the Sun pass through to the ground, where it turns into heat. When the ground radiates this heat back up as long-wave infrared rays, the carbon dioxide traps it. This dense heat trap keeps Venus hot even at night.

The atmospheric pressure on the surface of Venus is more than 90 times our own. (That much pressure is what you would feel nearly a kilometer down in the ocean.) We walk through the dense atmosphere in slow motion. The distant landscape shimmers. Only those objects nearby stand out in sharp outline, since we cannot see things clearly for more than 100 meters—about the length of a football field.

Where does the Sun shine?

A huge hollow in the clouds over Venus's north pole may be 1,000 kilometers wide. Perhaps the Sun shines through this window. Also, shadows in the Venera photographs suggest that there may at least be a hazy Sun shining over Venus's surface.

One big question is how Venus became a hothouse planet. Some scientists speculate that Venus once was more like Earth than it is now. Maybe it had an atmosphere something like our own. Maybe it also had rivers and oceans.

Possibly a runaway greenhouse effect took over, driving the temperature so high that all the surface water boiled away. A lot of water vapor in the air would become an additional heat trap and send the temperature even higher. Eventually it would be hot enough to free large amounts of carbon dioxide, perhaps locked up chemically in the rocks. And so the planet would slowly build up its atmosphere of carbon dioxide. If this theory is true, what became of the water vapor that began the process? Some scientists think it broke up into hydrogen, which escaped into space, and oxygen, which was trapped in the rocks of Venus's crust.

Venus's average orbital speed of 126,000 kilometers an hour takes it once around the Sun every 224.7 Earth days. This makes it the second fastest planet in the Solar System—because it is the second closest to the Sun. Since a Venus year is much shorter than an Earth year, a 15-year-old by Earth time would be 24 years old on Venus and could expect to live to the age of 122.

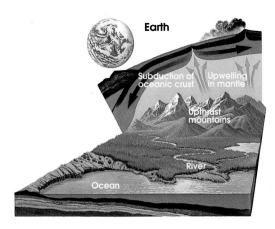

Earth

Subduction of oceanic crust
Upwelling in mantle
Upthrust mountains
River
Ocean

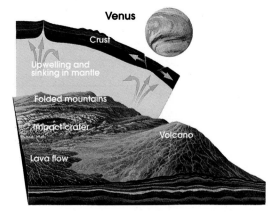

Venus

Crust
Upwelling and sinking in mantle
Folded mountains
Impact crater
Volcano
Lava flow

Volcanism and tectonics have shaped both Earth and Venus but Earth's landscapes are eroded by wind and water, and overgrown with plants. On lifeless Venus, volcanic forms are preserved. Scientists who study Venus can learn how Earth's mountains might appear without erosion.

The longest day

A Venus day would seem endless to us. In fact a Venus day is longer than a Venus year. The planet rotates on its axis once every 243 Earth days with respect to the stars. That is, every 243 days we would see the same stars in the same positions in the sky from Venus, compared with every 24 hours from Earth. But in the meantime, we would have watched two sunrises and two sunsets because Venus has moved along its orbit. All of this movement makes a Venus day—the period of rotation with respect to the Sun—117 Earth days long. From sunrise to sunset on Venus would last 117 times longer than on Earth. And since Venus rotates retrograde we would see the Sun rise in the west and set in the east.

A sunset once every 117 days would not only be a rare event but a spectacular one. Through the dense atmosphere, because of refracted light, we would see the Sun bent all out of shape, greatly flattened and stretched out along the horizon.

Parquet and ovoids

In the 1960's scientists saw Venus's surface with radio telescopes at the Goldstone Deep Space Network Tracking Station in California, and the National Astronomy and Ionosphere Center at Arecibo, Puerto Rico. They bounced radar waves off the planet's surface and picked up the reflected signals to put images together. These images revealed the presence of a few impact craters and large faults like those in East Africa, where crust is being pulled apart. And in a mountainous area called Ishtar Terra, about as big as the continental

United States, they showed terrain similar to that which forms on Earth when continents and plates crash together. However, Earth-based observations are limited because the same side of Venus faces Earth each time the planets are closest together.

In December 1978 two NASA Pioneer Venus spacecraft reached the planet. Pioneer Venus 2 dropped four probes into the atmosphere. Pioneer Venus 1, an orbiter, discovered many features that Earth-based radar could not detect.

In October 1983 two Soviet spacecraft, Venera 15 and 16, went into orbit around Venus. The radar images they returned of the northern part of Venus showed surface features as small as one or two kilometers.

They also revealed two types of terrain that are different from any we see on Earth. Soviet scientists called the first one "parquet" because it resembles the intricate patterns of inlaid wooden floors. The second type of feature is an ovoid, or corona—so-called because they have circular to oval, or crownlike, shapes. They look something like craters, but they are huge—up to hundreds of kilometers in diameter. Ovoids have not been found on any other planet or satellite in the Solar System. They are caused by the pressure of upwelling molten rock from Venus's interior.

In September 1990 NASA's Magellan spacecraft began radar mapping of Venus. Magellan's high-resolution images have shown us 99 percent of the planet's surface

with a clarity and detail that made its mission truly historic.

A new Venus

Aphrodite Terra—a continent-like region about the size of Africa—has hills and valleys similar to the Basin and Range area of Nevada and Utah, formed by stretching and collapse of the surface. In the north, in Ishtar Terra, the eastern slope of Freyja Montes has undergone uplift and compression but is now breaking up as parts of the crust collapse under their own weight. Western Ishtar Terra has a series of mountain ranges that are similar to the Rockies, the Andes, and the Appalachians. One of the ranges, Maxwell Montes, is 11 kilometers high, higher than Mount Everest. These mountains have folds and faults comparable to those of the Himalaya.

Scientists once asked: Are Venus's features shifting on giant crustal plates as Earth's are? Magellan seems to have answered the question. On Earth, tectonics involves horizontal movements of plates. On Venus, the movement is thought to be mostly vertical. Highland regions arise over mantle upwellings. Sinking areas may form mountain belts because of thickening and compression of the crust. Venus has hot-spot volcanoes similar to those that formed the Hawaiian Islands.

Thanks to Magellan, today we understand better than ever our mysterious neighbor in space.

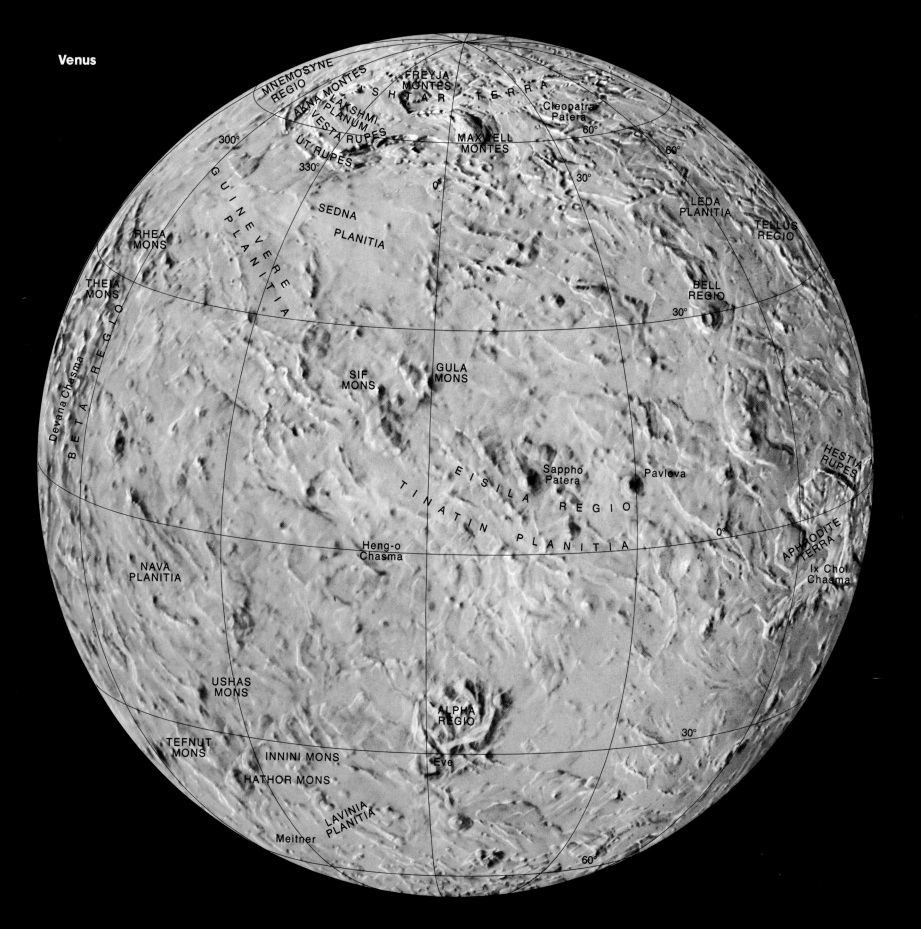

Venus

MNEMOSYNE
REGIO
AKNA MONTES
LAKSHMI
PLANUM
VESTA RUPES
UT RUPES

FREYJA
MONTES

I S H T A R   T E R R A

Cleopatra
Patera

MAXWELL
MONTES

300°

330°

0°

30°

60°

60°

GUINEVERE PLANITIA

SEDNA

PLANITIA

LEDA
PLANITIA

TELLUS
REGIO

RHEA
MONS

THEIA
MONS

BELL
REGIO

30°

BETA   REGIO

Devana Chasma

SIF
MONS

GULA
MONS

30°

HESTIA
RUPES

EISILA   REGIO

Sappho
Patera

Pavlova

TINATIN   PLANITIA

0°

NAVA
PLANITIA

Heng-o
Chasma

APHRODITE
TERRA

Ix Chol
Chasma

USHAS
MONS

ALPHA
REGIO

30°

TEFNUT
MONS

INNINI MONS

Eve

HATHOR MONS

LAVINIA
PLANITIA

Meitner

60°

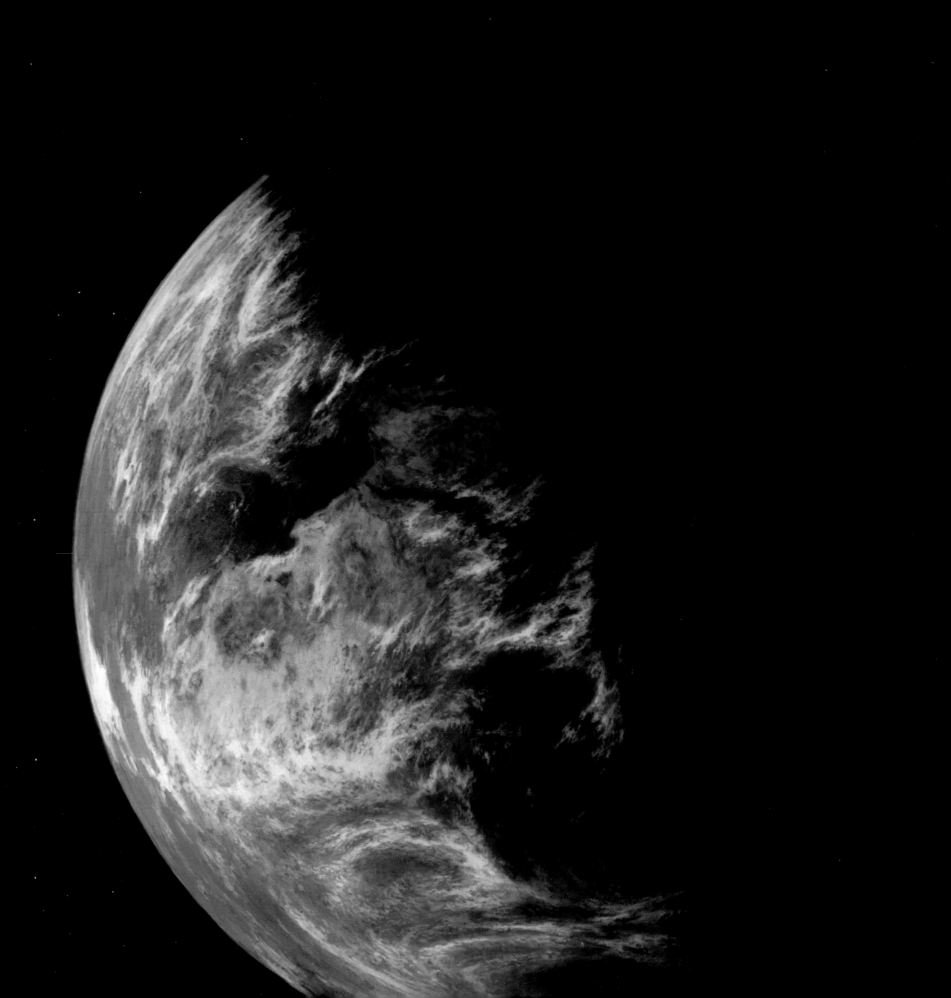

Earth

Delicate feather-like clouds that have formed from water vapor drift over and partly blanket Earth's surface—a surface mostly of water. So Earth deserves to be called "the watery planet."

With its one satellite, the Moon, Earth makes a breathtaking sight as we might see it from an orbiting spacecraft thousands of kilometers away. The Sahara sprawls across North Africa. You can locate the Mediterranean Sea to the north, with the Red Sea and the Arabian Peninsula to the east and, to the west, part of Spain and the blue of the Atlantic Ocean. But Earth's surface is never quite the same from one age to the next. Vulcanism and earthquakes change it, sometimes violently. Shifting sands and flowing water also bring changes. The forces of erosion hide or erase our planet's past.

But the Moon has no atmosphere, no water to alter its face. So the surface we gaze on today is largely a relic of geologically ancient times. Here on the Moon we can see the giant rayed crater Copernicus midway between the terminator and the limb. To the left of Copernicus is another big crater, Kepler, and Oceanus Procellarum, the Ocean of Storms. On the Moon, though, as we will see, oceans are not oceans at all.

Gaea, Earth goddess of the ancient Greeks, was called *Terra Mater* — Earth Mother — by the Romans. Symbol: ⊕ Greek sign for *sphaira*, sphere.

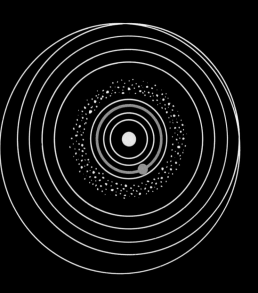

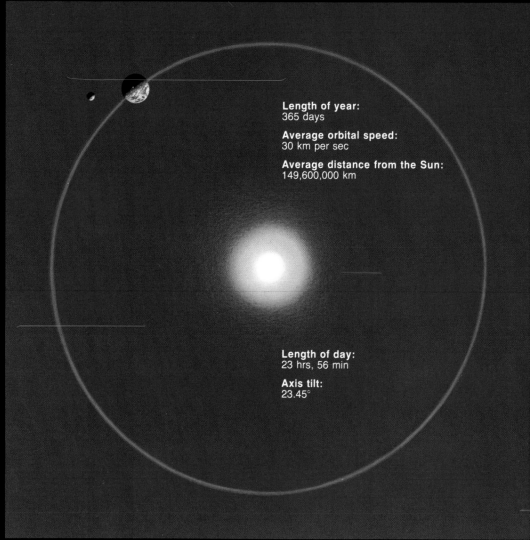

Length of year:
365 days

Average orbital speed:
30 km per sec

Average distance from the Sun:
149,600,000 km

Length of day:
23 hrs, 56 min

Axis tilt:
23.45°

Earth and its satellite, the Moon, follow a slightly oval-shaped path around the Sun. This causes our planet at one point to travel 2,500,000 km farther from the Sun than its average distance. Even so, Earth stays within a region of tolerance called the "ecosphere." In this safety zone, which roughly extends from the orbit of Venus to that of Mars, temperatures never get too high or too low to support our varied forms of life. If our orbit changed and carried us closer to the Sun, we would sizzle. Swinging too far away from the Sun, we would freeze into a ball of ice.

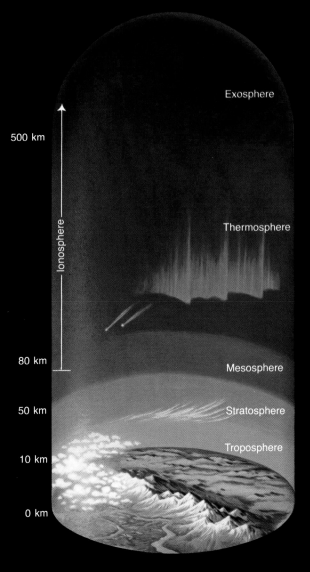

500 km

Exosphere

Ionosphere

Thermosphere

80 km

Mesosphere

50 km

Stratosphere

10 km

Troposphere

0 km

Earth's tilt causes the seasons. When the North Pole slants toward the Sun, the Northern Hemisphere has summer. Although Earth is then farthest from the Sun, the sunlight is less slanted and lasts longer. Winter comes to the hemisphere when the North Pole tilts away from the Sun.

Winter

Fall

Spring

Summer

Our layered atmosphere presses down on Earth. Most of our weather occurs in the troposphere. Air, mainly nitrogen and oxygen, also contains small amounts of water vapor and other gases. Meteors blaze into the atmosphere, while a curtain-shaped aurora glows brightly.

A magnetic field acts as our shield against the solar wind, creating a region called the magnetosphere. The wind carries deadly, electrically charged particles as it streams outward from the Sun. Some particles are trapped in the Van Allen Belts, two bands that circle Earth.

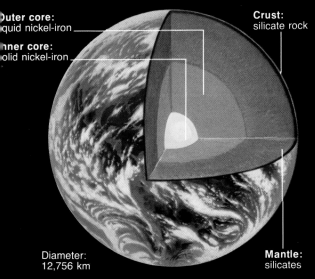

Outer core: liquid nickel-iron

Inner core: solid nickel-iron

Crust: silicate rock

Mantle: silicates

Diameter: 12,756 km

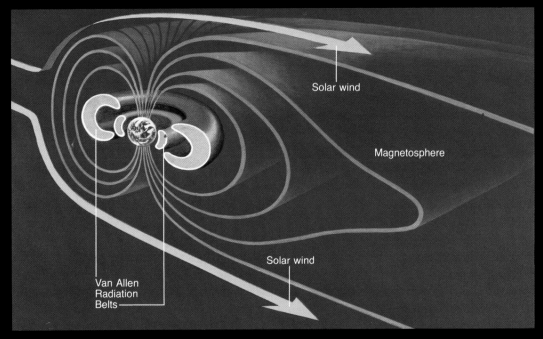

Solar wind

Magnetosphere

Van Allen Radiation Belts

Solar wind

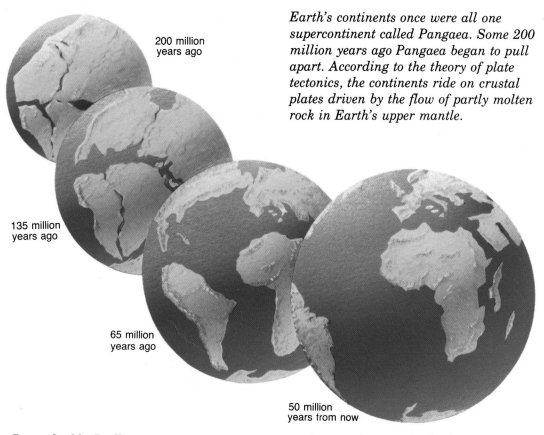

200 million
years ago

135 million
years ago

65 million
years ago

50 million
years from now

Earth's continents once were all one supercontinent called Pangaea. Some 200 million years ago Pangaea began to pull apart. According to the theory of plate tectonics, the continents ride on crustal plates driven by the flow of partly molten rock in Earth's upper mantle.

This NASA photograph from 756 km up shows how the Arabian Peninsula has torn away from Africa, creating the Red Sea (left) and the Gulf of Aden (right). Further drifting in ages to come will widen the Red Sea. The Horn of Africa may break off to become an island (left).

From dust to Earth

Within a spinning cloud of gas and dust tiny grains of matter collide and grow into larger and larger clumps.... The clumps sweep each other up and form a ball, producing enormous heat.... This is how scientists think Earth may have formed about four and a half billion years ago. Inside an already hot Earth, temperatures rose further as decaying radioactive elements released heat, until the ball became a soupy mass simmering at 2,000°C. While heavier matter, such as iron, sank toward the center and formed a super-dense core, lighter-weight crystals floated up toward the surface and began to form a crust.

Fissures cracked the hardening crust, and lava and trapped gases bubbled to the surface. They formed a primitive atmosphere, different from the one we know today. It was made up mostly of nitrogen, carbon dioxide, carbon monoxide, and water vapor with, possibly, small amounts of methane and ammonia. As the water vapor rose above the planet and cooled, it condensed and fell as rain. In some places the surface remained so hot that the rains kept evaporating back into water vapor as quickly as they fell.

As time passed, surface rock absorbed most of the first rains. When the crust could not sponge up any more, the waters slowly collected in vast basins and became our first seas. Through thinning clouds the Sun at last shone on a planet of jagged rock and sparkling water.

One continent

The continents we know today did not exist then. But continental masses did exist, and about 220 million years ago they had merged into the single supercontinent of Pangaea. By 135 million years ago, Pangaea had broken up and drifted apart into a northern half called Laurasia and a southern half called Gondwana. By 65 million years ago these two huge land masses had split further and the pieces had drifted toward the positions of the continents we know today.

The changing planet

The continents are still drifting. They rest on crustal plates that are moved by convection currents in the partially molten mantle, like large pieces of rock on conveyor belts. The energy that produces these motions is heat welling up from deep within our planet. Where the plate edges grind against or ride over each other, volcanoes sometimes erupt, earthquakes rumble, or an edge crumples up and forms mountains. When two plates beneath an ocean pull apart, the shape of the ocean floor gradually changes. The Atlantic Ocean floor spreads some four centimeters a year, while the Pacific Ocean floor shrinks.

World climate also changes, influenced by polar ice. Ice ages alternate with warm periods. Over the past two million years— the Great Ice Age—there may have been as many as 18 glaciations. At the peak of the last ice advance, some 18,000 years ago, about 30 percent of the land surface was under ice. In the polar regions, the ice seems permanent, but it is not. At times in Earth's past, the average global temperature was about 22°C, warm enough to keep even the poles free of ice. Today the average temperature, 14°C, keeps Greenland (once covered with forests) and Antarctica under ice caps many kilometers thick.

If a warm period melts the sheets of ice on Greenland and Antarctica, so much water will be released that the level of the

Molten rock burst through a new vent of Kilauea volcano on the island of Hawaii in 1977. Lava poured from the vent and down the mountain in a river of fire 10 m deep, 300 m wide, and nearly 10 km long. Kilauea, one of the world's most active volcanoes, erupts frequently.

Red lines show the boundaries of Earth's crustal plates. Along these seams—called lines of fire—the surface is weakest. Edges rub and override each other, causing earthquakes. Volcanoes are created when magma, under tremendous heat and pressure, oozes out through cracks.

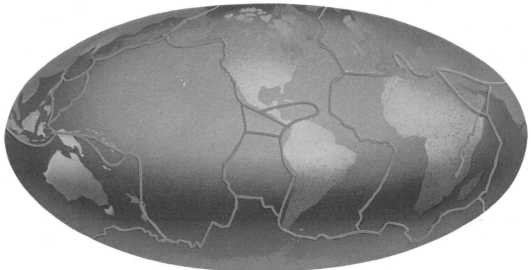

oceans will rise and drown coastal cities and towns around the world.

Mountain building

Earth is a living planet, its face sculptured by ice, wind, and water, and altered by the slow uplift of mountains. The Appalachian chain began to take shape over 400 million years ago. About 230 million years ago, these mountains south of New England crumpled up as the North American and African continents came together. The ancestral Sierra Nevada mountain range rose 140 million years ago, when molten rock pushed into the upper crust. The modern Sierra Nevada in California were thrust up in the last several million years. The earliest Rocky Mountains were formed about 300 million years ago. But the modern Rockies, which are still rising, were formed 70 to 100 million years ago.

In great outbursts of fire and lava about two million years ago, Mount Shasta was born in California and Mount Rainier in Washington. The eruption of Washington's Mount St. Helens, which blew its top in 1980, reminds us of how quickly Earth can change. But an example of slow change goes steadily on in Arizona. There, for millions of years, scouring wind, beating rain, rock-cracking frost, and the surging water of the Colorado River have been carving out the Grand Canyon.

Mountains under the sea

Covering most of our planet's surface, the oceans reach an average depth of 4.5 kilometers. But trenches in the ocean floor are twice that deep. The Mariana Trench in the Pacific, deepest point on Earth, plunges to 11 kilometers below sea level. It is deep enough to hold seven Grand Canyons stacked in a pile. Another gaping chasm in the Pacific, the Kermadec-Tonga Trench, is long enough to stretch from New York City to Kansas City.

More than four billion years ago gas and dust particles combined into the ball we call Earth. Intense internal heat turned solid matter into molten rock. Lava poured from fissures, cooled, and formed a thicker crust (upper). Locked inside the emerging lava were the elements hydrogen and oxygen in the form of water vapor. At the surface the water vapor, along with other gases, formed a primitive atmosphere. During Earth's first billion years, surface heat diminished. The water vapor condensed, and heavy rains fell. On hitting hot spots, the rain evaporated as steam (lower), then fell again. Eventually the surface cooled. Gradually the cracked crust absorbed water until underground spaces could hold no more. Surface water then began to collect in basins and depressions (opposite), and the first oceans were born.

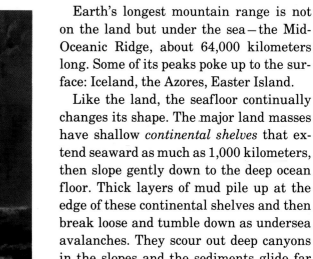

Earth's longest mountain range is not on the land but under the sea—the Mid-Oceanic Ridge, about 64,000 kilometers long. Some of its peaks poke up to the surface: Iceland, the Azores, Easter Island.

Like the land, the seafloor continually changes its shape. The major land masses have shallow *continental shelves* that extend seaward as much as 1,000 kilometers, then slope gently down to the deep ocean floor. Thick layers of mud pile up at the edge of these continental shelves and then break loose and tumble down as undersea avalanches. They scour out deep canyons in the slopes and the sediments glide far out onto the deep ocean floor.

Rivers in the sea

Blown by winds, the surface waters of the oceans flow around the continents. They move in currents like rivers within the ocean itself.

One of the best known of the warm currents is the Gulf Stream. It sweeps up the East Coast of the United States, then curves gently north and eastward past Britain toward Iceland.

Heat carried by the Gulf Stream warms northern lands. There also are cold currents, such as the Labrador Current that flows out of the Arctic near Greenland. There are deep, almost freezing, currents that inch northward along the sea floor from Antarctica. Just as air interacts with the land, it also interacts with the oceans, exchanging materials with them and driving their motion. The chief energy source for this activity—both on the land and within the sea—is the Sun.

Life on Earth 150 million years ago saw lush vegetation and many forms of reptiles. Two pterosaurs—reptiles, not birds—soar in the distance, while a pair of feathered, flightless Archaeopteryx— *the earliest known bird—team up to confront a* Compsognathus, *a small,* *bipedal dinosaur.* Archaeopteryx *shows links to reptiles (sharp teeth, claws, and a long, bony tail) and modern birds (wings and feathers). Because it lacked a strong breastbone to anchor muscles, the bird could probably only flap in pursuit of prey, using its wings like a net.*

Exploring Earth's interior

The center of our planet lies nearly 6,400 kilometers below the surface. In recent times, we have drilled down into the crust to learn more about Earth. So far we have reached a depth of only 12.3 kilometers. But geologists dream of drilling even deeper and bringing up a sample of material. That would tell us exactly what our planet's interior is made of. Since we cannot drill to the center, we must get our information in other ways. One important way is to study and measure earthquakes.

During an earthquake different kinds of shock waves, called *seismic* waves, shiver around and through the planet. Several kinds of waves travel away from the focus of the earthquake at different speeds. The primary waves are important to science because they travel all the way through the planet. Because we know how these waves behave as they pass through different materials, such as rocks and liquids, we can piece together the structure and composition of Earth's interior.

But to study the crust we don't have to wait for earthquakes. Geophysicists make their own mini-quakes by setting off explosives underground. These set up seismic waves that are recorded on instruments called *seismographs,* and we then have a meaningful record to read.

Seismic waves tell us the density of the rock they happen to be passing through. The waves travel through very dense rock, such as basalt, faster than they travel through less dense rock, such as sandstone. We also know that certain waves can travel through solids, liquids, or gases, while others, such as secondary waves, can pass only through solids—not through liquids. By recording the travel times and paths of seismic waves through Earth, scientists have drawn this profile:

There are three main layers—the crust, mantle, and core. The lightweight surface rocks of the *crust* are about 2.8 times denser than water. The crust extends downward to about 40 kilometers under the continents and to about 5 kilometers under the seafloor. The rocks of the crust are those familiar to us at Earth's surface. Although 40 kilometers sounds like a lot, in this case it is not. The crust makes up only one percent of Earth's volume. If Earth were an apple, the apple's thin skin would equal the thickness of the crust.

From high above Earth's "skin" a number of surface features can be seen in detail. Among them are impact craters made by meteorites. For instance, the remains of at least 17 craters are scattered over the Canadian Shield, a vast area of two-billion-year-old rock that cups Hudson Bay. The oldest parts of Earth's crust show the most craters. Younger parts of the crust formed after most meteorite activity was over. During the millions of years since the craters formed, the majority have been removed by erosion or buried under sedimentary rock. Early in Earth's history its surface probably was pitted everywhere with impact craters. Today we know of nearly 80, but undoubtedly there are many more. They range from less than 100 meters to more than 100 kilometers across. One, in Vredefort, South Africa, measures 140 kilometers in diameter.

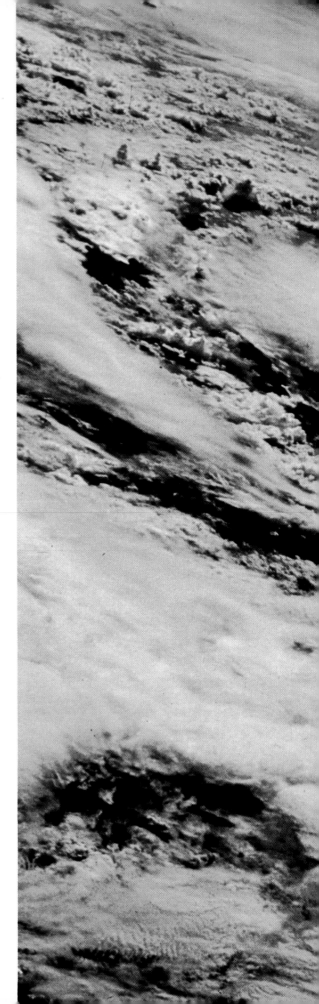

Hurricane Gladys, photographed by Apollo 7 astronauts, swirls over the Gulf of Mexico. The Coriolis effect, caused by Earth's rotation, governs the direction of spin in such storms. The movement is counterclockwise in the Northern Hemisphere, clockwise in the Southern.

Below the crust lies the *mantle,* a layer of rock that extends halfway to the center of Earth, or to a depth of 2,900 kilometers. Its upper part seems to be brittle rock, for earthquakes have been recorded as deep within the mantle as 700 kilometers. Parts of the mantle get so hot that the rock becomes molten and moves slowly in vertically rotating currents.

Earth's *core* has a fluid outer layer enclosing a solid inner region, both mostly of iron and a little nickel. The inner core is a ball about 2,600 kilometers across. Earth's temperature increases from about 15°C at the surface and may reach 4,000°C in the inner core. The pressure in the inner core is extremely high, more than three times that inside Earth's mantle, and over three million times that of our planet's atmosphere at the surface.

The marvel of life

Some 3.5 billion years ago, most scientists believe, bacteria and tiny plantlike organisms began to grow in the seas. Over millions of years some evolved into complex plants. They gradually developed a process called *photosynthesis.* By this process, plants containing a green substance, chlorophyll, use sunlight to make their own food. They take in carbon dioxide and water and change them into sugar, which they use as energy. As a result of the process, they give off oxygen and produce water.

These plants were so successful that they changed Earth's atmosphere in an important way. The oxygen they gave off first reacted with iron dissolved in sea water, causing it to oxidize, or rust, and settle on the ocean floor. When most of the iron was used up, the surplus oxygen accumulated to such an extent that it started escaping into the atmosphere, forming part of the air we breathe today.

As hundreds of millions of years passed, water plants took up life on land and spread far and wide. So did fish-like organisms that inhabited the seas in the Age of Fishes, 395 to 345 million years ago. Much later, during the Age of Reptiles, 225 to 65 million years ago, those magnificent "terrible lizards," the dinosaurs, ruled Earth for well over a hundred million years. When dinosaurs became extinct, some 65 million years ago, their place of dominance in the animal kingdom was taken by mammals, which later included humans. We still live in the Age of Mammals.

Today 1.4 million different kinds of animals and 500,000 different kinds of plants are known to inhabit Earth. If that seems a lot, then think of this: 99 percent of all the kinds of plants and animals that ever lived on Earth are extinct. Some very old types have escaped extinction by adapting to a changing environment. The cockroach and the shark were abundant 250 million years ago. The 19th-century naturalist Charles Darwin marveled at Earth's varied life when he wrote: "We may well affirm that every part of the world is habitable! Whether lakes of brine, or those subterranean ones hidden beneath volcanic mountains—warm mineral springs—the wide expanse and depths of the ocean—the upper regions of the atmosphere, and even the surface of perpetual snow—all support organic things."

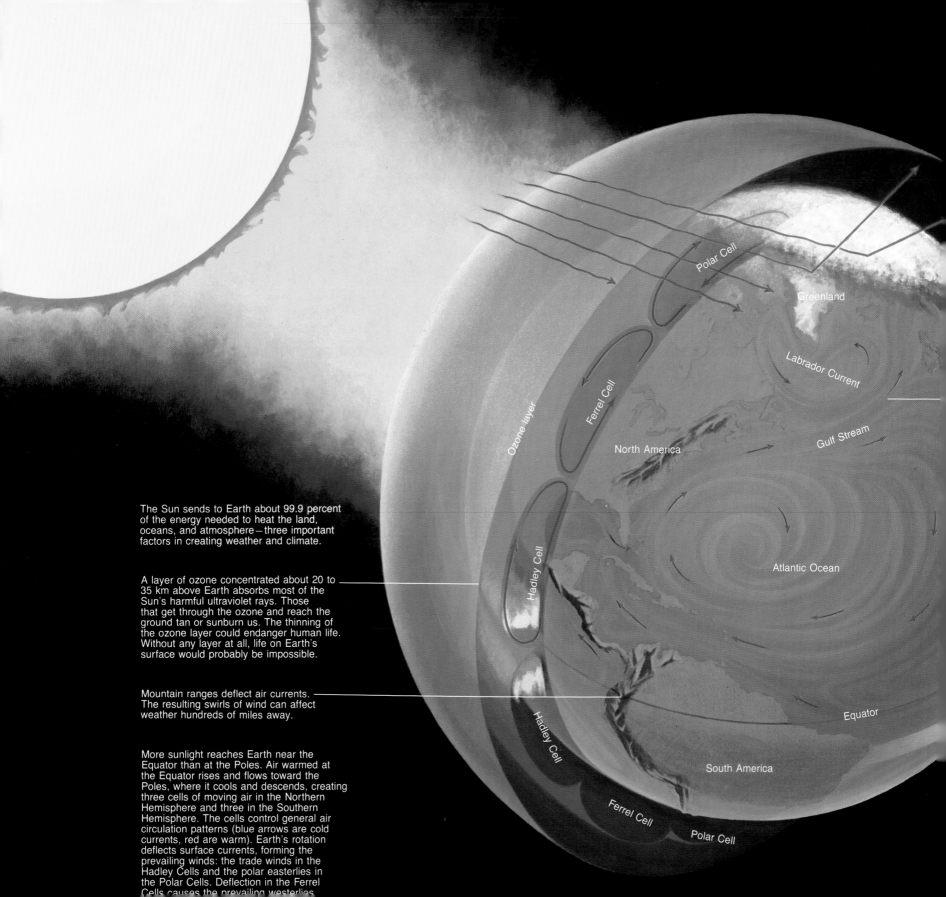

The Sun sends to Earth about 99.9 percent of the energy needed to heat the land, oceans, and atmosphere—three important factors in creating weather and climate.

A layer of ozone concentrated about 20 to 35 km above Earth absorbs most of the Sun's harmful ultraviolet rays. Those that get through the ozone and reach the ground tan or sunburn us. The thinning of the ozone layer could endanger human life. Without any layer at all, life on Earth's surface would probably be impossible.

Mountain ranges deflect air currents. The resulting swirls of wind can affect weather hundreds of miles away.

More sunlight reaches Earth near the Equator than at the Poles. Air warmed at the Equator rises and flows toward the Poles, where it cools and descends, creating three cells of moving air in the Northern Hemisphere and three in the Southern Hemisphere. The cells control general air circulation patterns (blue arrows are cold currents, red are warm). Earth's rotation deflects surface currents, forming the prevailing winds: the trade winds in the Hadley Cells and the polar easterlies in the Polar Cells. Deflection in the Ferrel Cells causes the prevailing westerlies.

Polar Cell

Greenland

Labrador Current

Ferrel Cell

Gulf Stream

Ozone layer

North America

Atlantic Ocean

Hadley Cell

Equator

Hadley Cell

South America

Ferrel Cell

Polar Cell

Solar rays reflected from snow, ice, and clouds affect the amount of heat that stays near Earth's surface. Open water, land, and the atmosphere absorb as well as reflect solar energy. Fine ash carried from erupting volcanoes to the upper atmosphere can cool Earth by reflecting some of the incoming solar heat.

Air pollution can cause global changes in our climate. When burned, fossil fuels like coal and oil release carbon dioxide (CO_2) and other chemical particles. These may form a blanket in the atmosphere, trapping surface heat — infrared rays — that normally would radiate back into space. This "greenhouse effect" can raise the overall temperature of the atmosphere.

rope

Earth's oceans store and transport heat. The warm Gulf Stream flows far north, keeping the climate of Europe moderate.

Africa

Desert winds raise dust and blow it long distances. Over the subtropics this dust sometimes blocks solar rays, causing temperatures to fall temporarily.

Earth's rotation

Solar energy regulates Earth's climate. Hot air rising at the Equator moves toward the poles. Cold polar air sinks and moves toward the Equator. Water evaporates; the vapor rises, then cools and falls as rain or snow. Earth's rotation and topography help establish wind and weather patterns.

Our ocean of air

On the average, Earth's air is a mixture of dust and gases — 78 percent nitrogen, 21 percent oxygen, and one percent other gases that include argon, neon, carbon dioxide, water vapor, and ozone. As we descend into the atmosphere from outer space, we pass through the *exosphere*. In this region the air is so thin and gravity so weak that atoms escape into space.

We next descend through four major layers. The first layer, which begins about 500 kilometers above the ground, is the *thermosphere*. Up here so few air molecules exist that the Sun's radiation has too little to strike against in order to scatter light. Thus, the sky is very dark. The thin atmosphere also means that there is no transfer of heat from the air to any object in the air. Any living creature exposed to the thermosphere would broil on its Sun side and freeze on its shadow side. At the bottom of the thermosphere, 80 kilometers above the surface, temperatures are –90°C.

Below the thermosphere is the *mesosphere* layer, about 30 kilometers deep. The sky here is nearly dark, since the air still is very thin. At the base of the mesosphere the temperature climbs to about 0°C. The *ionosphere* also begins about here. It makes long-distance radio communication possible. Radio waves sent from stations on the surface of Earth travel up to the ionosphere, glance off it and back down to radio receivers many miles away. Without the ionosphere, radio signals from our transmitters would pass right on out through the atmosphere.

The next layer down is the *stratosphere*,

On the next page: *Running water, like Iguazú Falls in Brazil, is one agent of erosion changing Earth's face. Wind, rain, freezing, and thawing also wear away the rock. Over millions of years, they may scour a mountain down to a plain or carve a ditch as big as Grand Canyon.*

about 40 kilometers deep. At the bottom of this layer the temperature dips to about –50°C, much colder than at the top. How can this be, since the air is supposed to get steadily warmer as we move deeper into the denser lower region of our ocean of air?

The answer is a strip in the middle of the stratosphere called the *ozone layer*. Ozone is a special form of oxygen with molecules that have three atoms instead of two. This layer of gas shields us from damaging ultraviolet energy entering the atmosphere from the Sun. Absorption of this energy causes the ozone layer to heat up.

The bottom air layer is called the *troposphere*. As we descend through it into denser and denser air, we become gradually warmer. The warming comes about when short-wavelength energy from the Sun heats up the ground and the oceans and is returned to the atmosphere as long-wavelength (heat) energy.

High in the troposphere we find ourselves battered by jet-stream winds blowing 400 kilometers an hour. As we continue down, the winds lessen and the sky gradually changes from deep to lighter blue. The sky is blue because the short, blue wavelengths of light get scattered more than the longer, red wavelengths.

In this blue layer of air — which reaches from the ground up to about 10 kilometers — all of Earth's weather takes place. Here we find the westerlies, the gentle trades, thunderstorms, and lightning.

Earth as a magnet

Earth behaves as though it had a giant bar magnet inside, running from north to

Earth

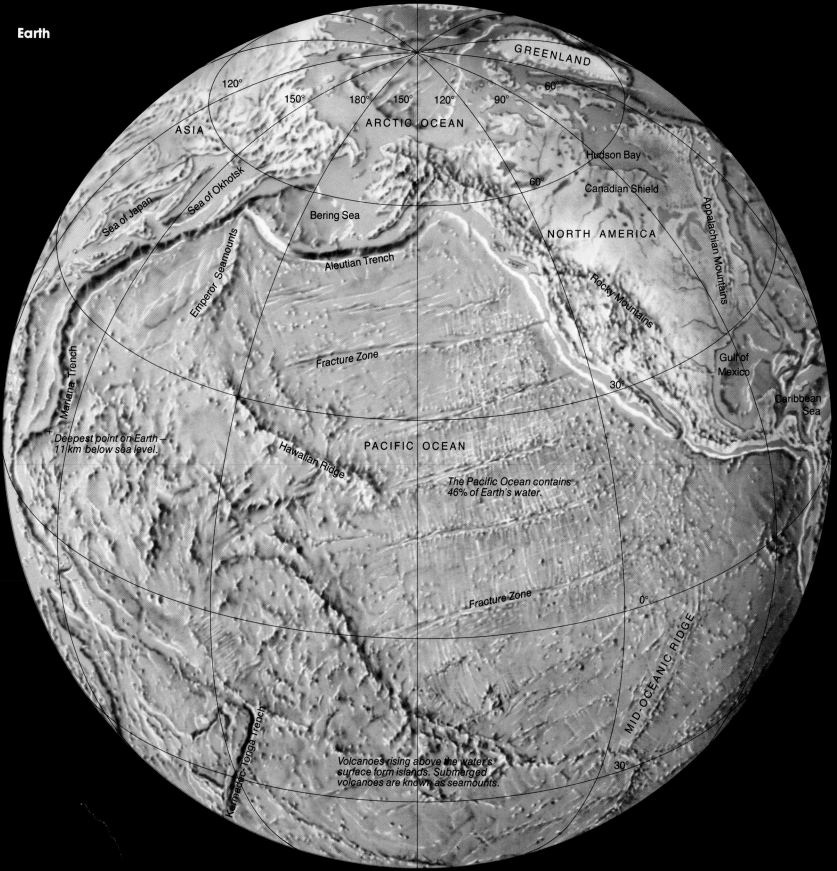

GREENLAND

120°
150°
180°
150°
120°
90°
60°

ARCTIC OCEAN

ASIA

Hudson Bay

60°

Canadian Shield

Sea of Japan

Sea of Okhotsk

Bering Sea

NORTH AMERICA

Appalachian Mountains

Emperor Seamounts

Aleutian Trench

Rocky Mountains

Mariana Trench

Fracture Zone

30°

Gulf of Mexico

Caribbean Sea

Deepest point on Earth —
11 km below sea level.

Hawaiian Ridge

PACIFIC OCEAN

The Pacific Ocean contains
46% of Earth's water.

Fracture Zone

0°

MID-OCEANIC RIDGE

Kermadec-Tonga Trench

30°

Volcanoes rising above the water's
surface form islands. Submerged
volcanoes are known as seamounts.

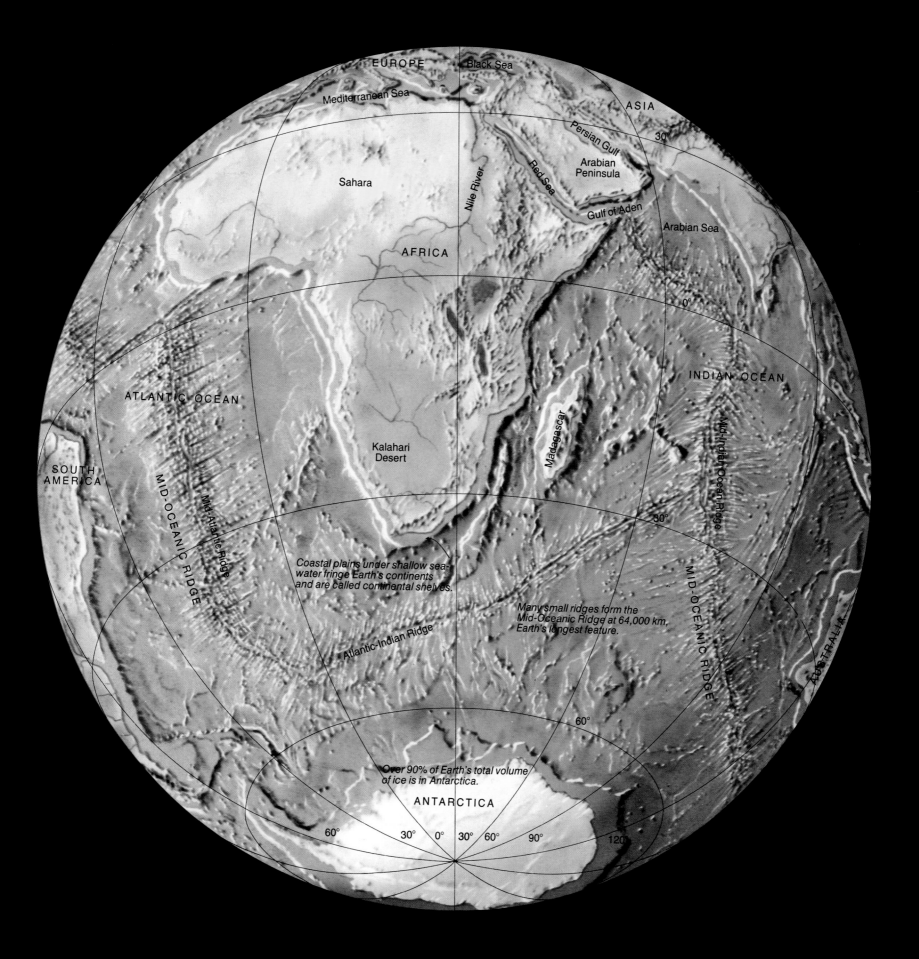

Earth's lowest tides occur when Moon, Earth, and Sun form right angles. When the bodies line up, their gravity causes the highest tides. On Earth's opposite side, centrifugal force pulls more strongly than the Moon's gravitational attraction, creating a bulge of water — high tides.

Under a full Moon, high tide creeps up the sand at Myrtle Beach, South Carolina. The Moon's gravitational attraction pulls at our oceans as Earth rotates, making most coastal waters rise and fall. About every 12.5 hours tides are highest (or lowest) on opposite sides of Earth.

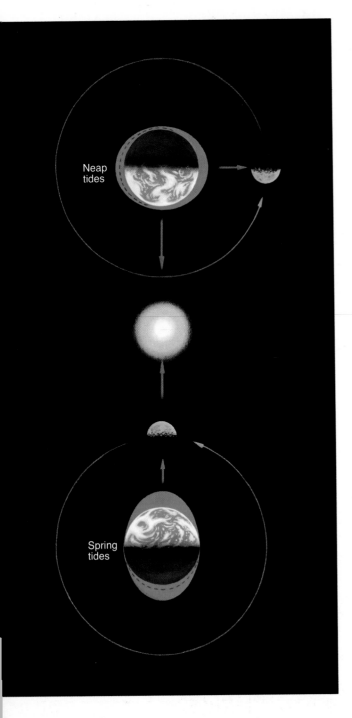

Neap tides

Spring tides

south. Heat flow within the molten outer core of Earth causes slow convection, or circulatory, currents in the liquid metal. This motion in the outer core is thought to generate lines of magnetism between the magnetic north and south poles, producing Earth's *magnetic field*. The solar wind squashes the magnetic field against Earth on the Sun side and stretches it out on the shadow side, creating the *magnetosphere*.

Earth's magnetic field protects us from dangerous solar radiation. Most of the charged atomic particles from the Sun are deflected around Earth by the magnetic field. But some are trapped — for a while — in a great doughnut-shaped "holding" area around Earth called the Van Allen radiation belts. The particles eventually leak off into space as new ones are taken captive. Under certain conditions, when solar wind particles strike the Van Allen belts they are speeded up. As they hit the upper atmosphere, they then glow as "northern lights," or the aurora borealis (aurora australis in the Southern Hemisphere).

Some radio waves — generated by lightning — are transmitted along the magnetic field lines between the poles. When they reach one pole, they are repelled and bounce back toward the opposite pole. Back and forth they go. During their journeys they make whistling sounds that we can hear by radio. A whistler makes one trip from pole to pole in about one second!

A watery planet

Water covers 70 percent of Earth's surface and is part of the atmosphere as well. Essential to life, water is so widespread that plants and animals are found almost everywhere. They help to recycle and renew the atmosphere. Plants take up water through their roots and return it to the air through their leaves as water vapor. When animals breathe, they exhale water vapor. You can see the moisture when you breathe on a window or mirror. When animals die, the water that makes up as much as 70 percent of their bodies is released and returned to the air or the ground.

Plants and animals are only a small part of the exchange process called the *hydrologic cycle,* or water cycle. Ocean water evaporates and changes into water vapor. Rising in the atmosphere, it cools and condenses into droplets. These collect in clouds, which return the water to the surface as rain and sometimes snow. Some of this *precipitation* is evaporated right away, some flows in rivers back to the ocean, and some seeps into the ground.

Ocean water is recycled another way. Some scientists say that, over the course of six to eight million years, water equal to the volume of all the oceans enters cracks where plates pull apart the ocean floor. Deep below, the water heats up, changes its chemistry, and reappears along with welling lava.

Tides and the Moon

Earth has one natural satellite, the Moon. It is an airless world with a diameter almost one-fourth that of Earth. In our Solar System only Pluto, with its companion Charon, has a satellite so close to its own size. We look on the Earth-Moon system as a double planet. *(Continued on page 116)*

Diana, or Luna, was the Roman goddess of the Moon, animals, and hunting. From Latin *lucere*, to shine, Luna gives us "lunar." Symbol: ☾ a crescent Moon.

Earth

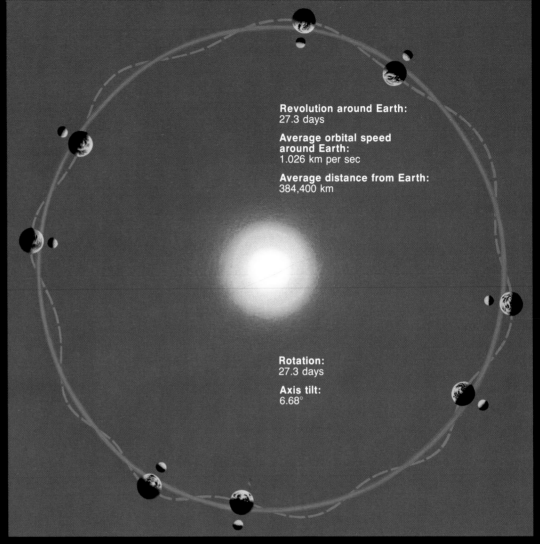

Revolution around Earth:
27.3 days

Average orbital speed around Earth:
1.026 km per sec

Average distance from Earth:
384,400 km

Rotation:
27.3 days

Axis tilt:
6.68°

Moon and Earth, tied together by gravitational attraction, revolve as a double planet. Think of them as unequal ends of a weight-lifter's barbell. Because Earth's mass is 81 times greater than the Moon's, the center of gravity of the Earth-Moon barbell lies about 1,700 km below Earth's surface. It is called the *barycenter*. This pivot point—not Earth's geographical center—follows the smooth orbital line in the diagram. As the barbell spins around its eccentric center of gravity, both Earth and the Moon trace wobbly orbital paths through space.

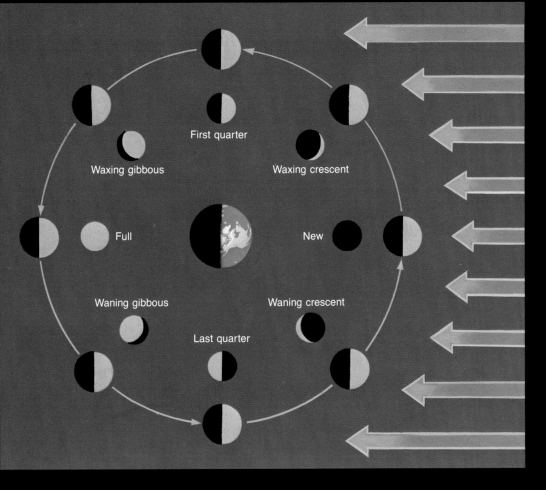

First quarter

Waxing gibbous

Waxing crescent

Full

New

Waning gibbous

Waning crescent

Last quarter

Total eclipses occur when the Moon and Earth line up perfectly with the Sun. During a solar eclipse, the Moon passes between the Sun and Earth, hiding the Sun on a small area of Earth (A). During a lunar eclipse, the full Moon moves into Earth's shadow and is blacked out (B).

On its trip around Earth, the Moon passes through phases of reflected sunlight. The inner circle above shows the phases we see from Earth. The outer circle shows the Moon as seen from high above our North Pole. Because Earth spins faster than the Moon revolves, the

Moon rises an average of 50 minutes later each night. During the new phase, Moon and Sun rise and set at the same time. From then on, the Moon appears in different parts of the sky: in the west as it waxes larger towards gibbous (hump shaped), in the east as it wanes smaller.

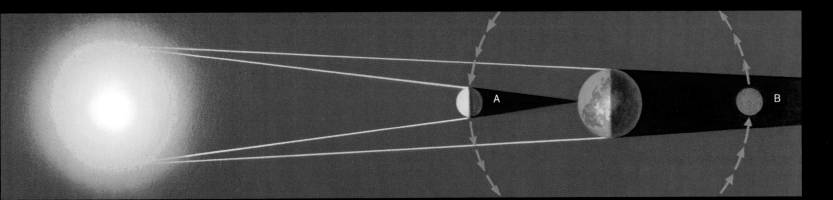

The gravitational force of the Moon, and to a lesser extent the Sun, raises the ocean tides on Earth. A great bulge of water is pulled up on the side of Earth facing the Moon and held there. As Earth rotates beneath the bulge, *high tides* occur. Because of Earth's rotation, the tide seems to move from east to west. If there were only the one bulge, we would have only one high tide a day. But there are two tidal bulges that cause two high tides each day—one every 12 hours and 25 minutes, each followed by a *low tide*. What causes the second tidal bulge?

Gravitation, remember, weakens as distance becomes greater. So the Moon tugs with greater force on that side of Earth facing the Moon, with less force on matter at Earth's center, and with the least force on Earth's far side. Because, on our globe's far side, the outward pull of *centrifugal force* is stronger than the inward pull of the Moon's gravitation, another bulge of water forms—our second high tide.

The Moon also raises tides in Earth's atmosphere and in the ground beneath our feet. In fact, the Moon tugs on every object on Earth's surface—including you. The ground tides amount to only a slight bulge spread across a great expanse of surface, while the highest water tides can reach 16 meters. As Earth rotates beneath its tidal bulges, friction gradually slows it down, by a tiny fraction of a second each century. This is not very much in a human lifetime, but in Earth's lifetime of several billion years it adds up. In 100 million years from now an Earth day will be almost half an hour longer than it is now.

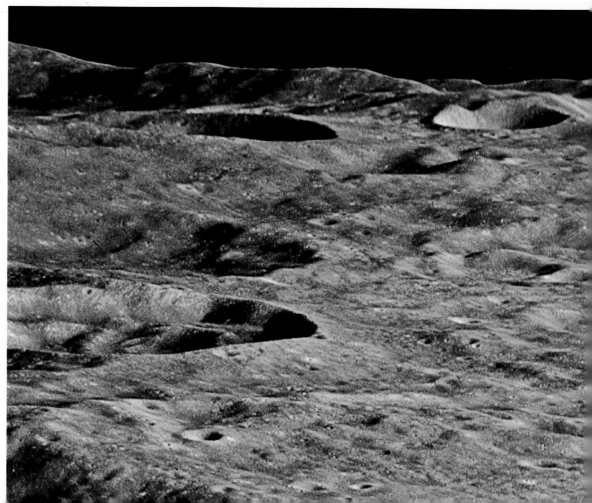

Far side of the Moon (lower), first seen in photographs taken by unmanned space vehicles, reveals a surface heavily pitted with craters caused by the impact of meteorites. It has a few of the lava-filled basins called seas, or maria, apparently because the far side has a thicker crust.

As Earth's rotation slows down, the loss of speed causes the Moon to spiral outward from Earth about three centimeters more each year. A hundred million years from now our Moon will be almost 3,000 kilometers farther away.

Whenever the Sun, Moon, and Earth are lined up, the Sun's gravitational force added to that of the Moon causes especially strong tides. We call these *spring tides* (but they are not related to the seasons). They happen twice a month, once at new Moon and again at full Moon. When the Sun, Moon, and Earth form a right angle with Earth at the corner, we have especially weak tides called *neap tides*.

Where did the Moon come from?

Some scientists once thought that long ago Earth captured a small planet—and that planet is the Moon. Although this seems to be how the giant planets obtained some of their satellites, most astronomers today do not accept the capture theory for the Moon.

According to another theory, the Moon and Earth were formed at the same time, out of the same gas and dust. The same elements are found on both—calcium, aluminum, titanium, magnesium, silicon, oxygen, iron—but in such different proportions as to cast serious doubt on this theory. There is very little carbon on the Moon. And the infant Moon did not pick up as much iron. That helps explain its low density—only 60 percent of Earth's.

Yet another theory, first proposed in the 19th century, suggested that the infant Earth was a rapidly rotating molten body that split in two. The Moon was the smaller of these two bodies. A different version of this theory held that the Moon was formed from a ring of fragments that had evaporated or been detached from the hot, primitive Earth.

Today, because of what we have learned from the Apollo program, scientists generally favor the theory that an object about the size of Mars collided with the primordial Earth. This violent impact threw a cloud of fragments into orbit around the Earth. They eventually combined to form the Moon.

The oldest dated rocks from the Moon, brought back by astronauts, are about 4.6 billion years old. So the Moon is about the same age as Earth.

Shaping the Moon's surface

Early in its life the Moon melted to a depth of several hundred kilometers. Later the surface cooled and formed a rocky crust. For a long time huge meteorites bombarded the surface and left hundreds of thousands of impact craters. Some are gigantic. Copernicus crater measures 91 kilometers from rim to rim, and Tycho is 87 kilometers across.

Some of the craters have central peaks and circular mountainous walls about six kilometers high. Other craters measure only a few meters across. And rock samples show tiny craters made by high-speed dust grains from space. They are so small that a hundred of them would fit on your little fingernail.

The Moon has seas, but they are not seas of water. On the Moon a sea is called a *mare*, Latin for "sea." *Maria* are seas of

Split Rock dwarfs U. S. astronaut Harrison Schmitt east of Mare Serenitatis on the Moon's near side. Schmitt holds a "gnomon," an instrument which the astronauts used to measure scale and color for their surface photography. This December 1972 trip was the final Apollo mission.

hardened lava. They are found in basins that take the form of either irregular depressions or deep impact craters. From 3.7 to 3 million years ago, many maria were formed when molten rock rich in iron and magnesium broke through the floors of the basins and spread over the surface. Some lunar seas are as much as 1,000 kilometers across—one-third the width of the contiguous United States. Most maria are found on the Moon's near side.

Mountain chains, like the lunar Apennines (Montes Apenninus), sometimes border the maria for hundreds of kilometers and rise five kilometers. But unlike Earth mountains, which were formed during thousands of years, our satellite's mountains were thrust up instantly by the impact of asteroids or meteorites.

Because the Moon lacks air, it is a hot and cold little world like Mercury. The lunar high noon temperature reaches 134°C. On the Moon's night side the temperature drops to about –170°C.

Moonquakes and motions

Our Moon is not geologically dead. Because the Moon is so close to us, Earth's gravitational force has enough strength to raise ground tides there. These tidal forces may touch off moonquakes. There can be as many as 3,000 a year, but generally the interior of the Moon is much quieter than the interior of Earth.

In its youth the Moon stopped rotating in relation to Earth. Now, with one ground tidal bulge facing Earth, and another on the opposite side, the Moon appears "frozen" with the same face always toward us.

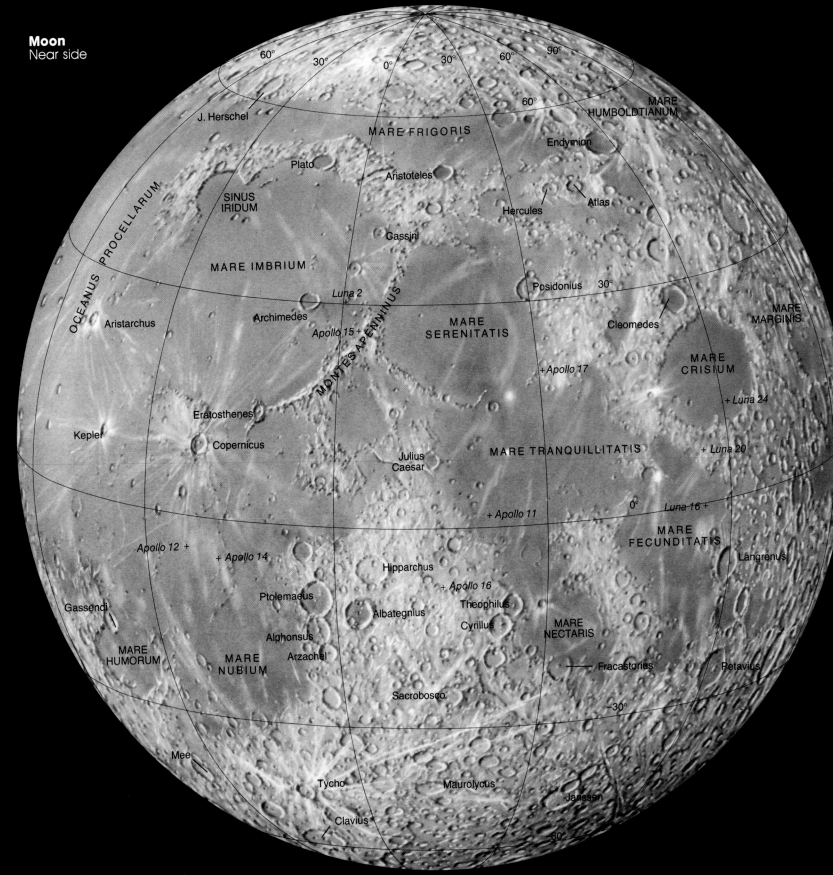

Moon
Near side

60° 30° 0° 30° 60° 90°

60°

J. Herschel

MARE HUMBOLDTIANUM

MARE FRIGORIS

Endymion

Plato

Aristoteles

SINUS IRIDUM

Hercules   Atlas

Cassini

OCEANUS PROCELLARUM

MARE IMBRIUM

Posidonius   30°

Luna 2

Archimedes

Aristarchus

MARE SERENITATIS

Cleomedes

MARE MARGINIS

Apollo 15 +

MONTES APENNINUS

+ Apollo 17

MARE CRISIUM

Eratosthenes

+ Luna 24

Kepler

Copernicus

Julius Caesar

MARE TRANQUILLITATIS

+ Luna 20

0°   Luna 16 +

+ Apollo 11

MARE FECUNDITATIS

Apollo 12 +

Hipparchus

Langrenus

+ Apollo 14

+ Apollo 16

Gassendi

Ptolemaeus

Theophilus

Albategnius

Cyrillus

MARE NECTARIS

Alphonsus

MARE HUMORUM

Arzachel

MARE NUBIUM

Fracastorius

Petavius

Sacrobosco

−30°

Mee

Tycho

Maurolycus

Janssen

Clavius

60°

120

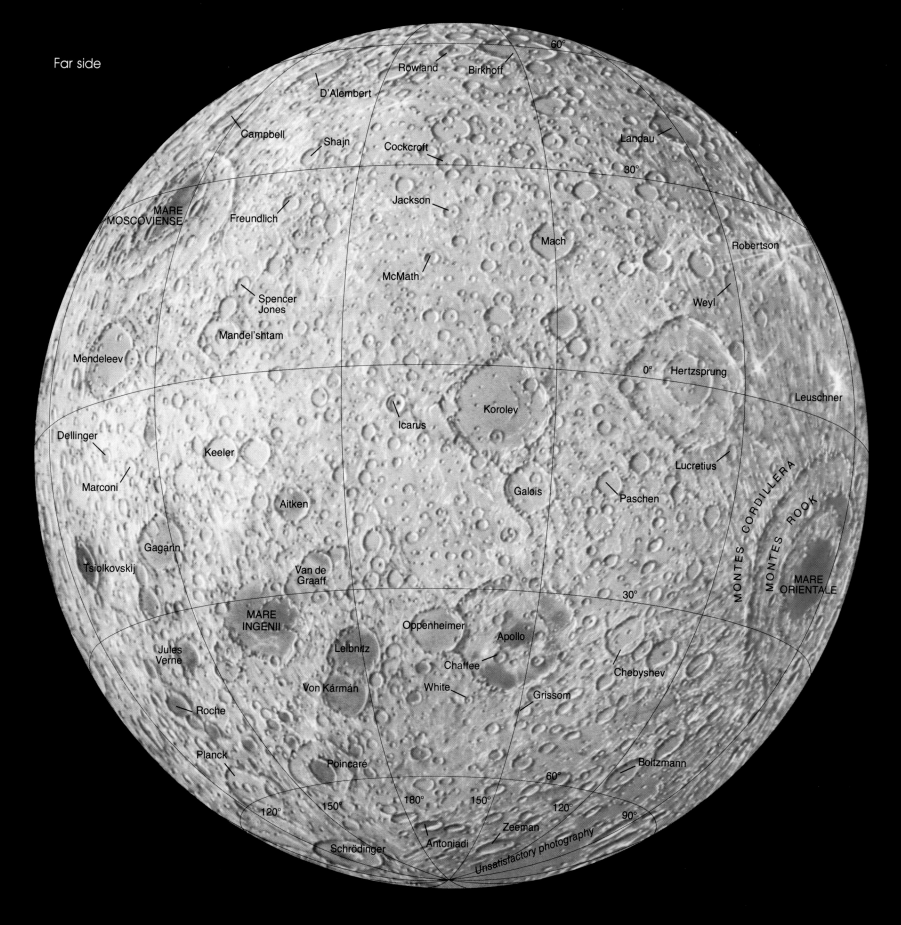

Far side

D'Alembert
Rowland
Birkhoff
60°
Campbell
Shajn
Cockcroft
Landau
80°
MARE MOSCOVIENSE
Freundlich
Jackson
Mach
Robertson
McMath
Weyl
Spencer Jones
Mandel'shtam
Mendeleev
0°
Hertzsprung
Leuschner
Dellinger
Keeler
Icarus
Korolev
Marconi
Galois
Lucretius
Aitken
Paschen
MONTES CORDILLERA
Gagarin
MONTES ROOK
Tsiolkovskij
Van de Graaff
30°
MARE ORIENTALE
MARE INGENII
Oppenheimer
Apollo
Chebyshev
Jules Verne
Leibnitz
Chaffee
Von Kármán
White
Grissom
Roche
Planck
Poincaré
Boltzmann
60°
120°
150°
180°
150°
120°
90°
Zeeman
Schrödinger
Antoniadi
Unsatisfactory photography

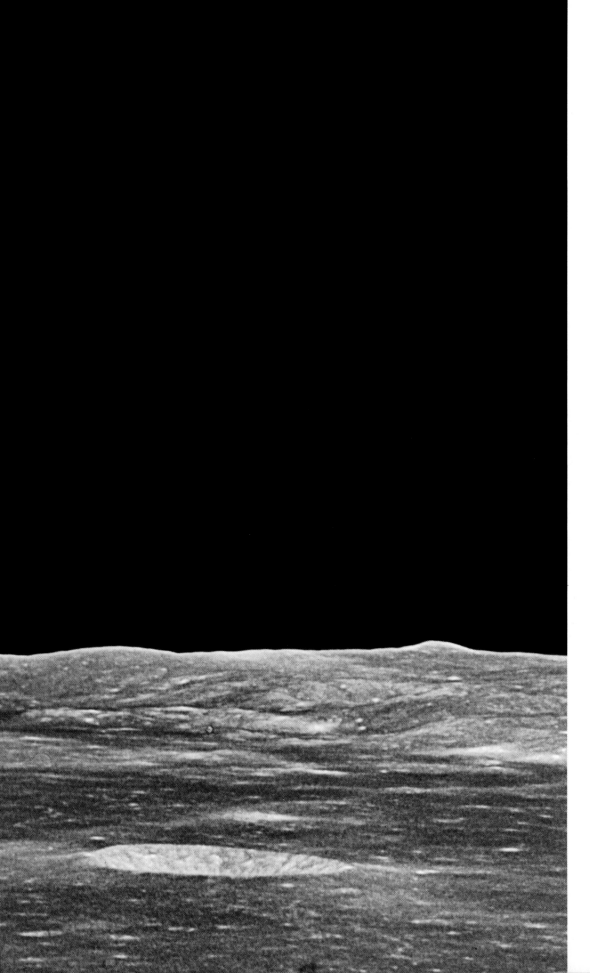

But from space, we would see the Moon complete one rotation every 27.3 days.

Men on the Moon

On July 20, 1969, humans first set foot on the Moon during the flight of Apollo 11. Leaving Earth on July 16, Neil Armstrong, Michael Collins, and Edwin Aldrin, Jr., reached their destination on the near side of the Moon and set up scientific experiments. Their mission completed, they returned to Earth on July 24 with the first Moon rocks. Theirs was the first of six United States lunar landings of the Apollo program, which ended in 1972 with Apollo 17. Eugene Cernan and Harrison Schmitt, the first geologist-astronaut to reach the Moon, landed December 11 near the Sea of Serenity (Mare Serenitatis).

Along with several unmanned Soviet Luna landings beginning in 1966, the Apollo explorations solved some of the mysteries about the Moon but also raised some questions. There is no life, but there are substances that may be the forerunners of amino acids—the building blocks of life. Nothing exists that can be called an atmosphere, but tiny amounts of gases—argon, neon—are found above the surface.

The Moon lacks a magnetic field and does not have magnetic poles, yet surface rocks are weakly magnetized. Were they magnetized by a core of liquid iron, once strongly magnetic? From its volcanic maria we know the Moon once baked with internal heat, and that it has been cooling for ages. But beneath the Moon's rocky mantle, there may still be a small, weakly magnetic core with some molten iron.

123

Mars

Three enormous volcanic mountains line up northeast to southwest—the Tharsis Montes. Each one is about twice as broad and over twice as high as the volcanic island of Hawaii measured from the seafloor. Almost hidden in shadow is Olympus Mons, an even larger volcano. Above it, wisps of water ice clouds hover. Farther north, the pole displays a shrinking cap of carbon dioxide snow typical of early spring.

East of the Tharsis Montes, a system of giant canyons stretches some 5,000 kilometers east to west. These canyons, as well as dried-up stream channels, many craters, and the great volcanoes, show that Mars had a very active geological past. But we don't know what tore open the giant canyons or created the channels.

We do know that the red color of Mars comes from a rustlike coating on the surface soil which sometimes is picked up and blown furiously in great storms. Here we see the Red Planet from its outer moon, Deimos, in orbit around Mars some 20,000 kilometers away. At its widest—15 kilometers—Deimos is only about one-fortieth as broad as Olympus Mons. It is heavily scarred with craters. So is Mars' other, larger moon, Phobos, the small object to the right of the planet's lighted limb.

Facts about
Mars

Like a badge of blood in the sky, the Red Planet has long stood for gods of war. Mars was the Roman war god. His shield and spear form the planet's symbol: ♂

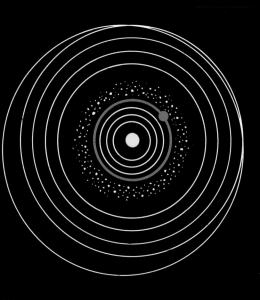

Length of year:
687 Earth days

Average orbital speed:
24 km per sec

Average distance from the Sun:
227,900,000 km

Length of day:
24 hrs, 37 min

Axis tilt:
25.2°

An off-center orbit: At one extreme, Mars loops 42.4 million km farther from the Sun than at the other. If Venus swung out that far, it would cross our own path. As on Earth, Mars' tipped axis causes seasons. Because Mars moves fast when close to the Sun, slower when far away, seasons differ in length. Northern spring lasts 52 Martian days more than fall. Two miniature moons zip around Mars. Deimos, 20,123 km high in the sky, orbits every 30 hours. Phobos, only 5,973 km up, takes less than 8 hours. Phobos rises and sets twice a day.

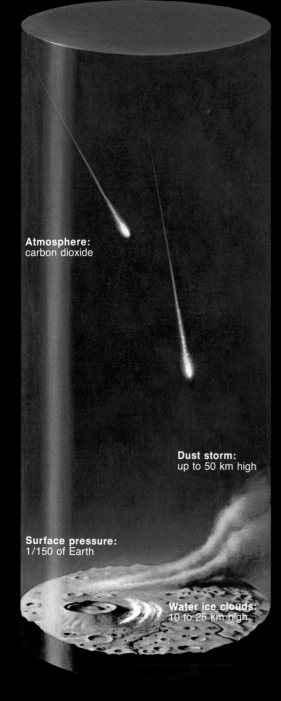

Atmosphere:
carbon dioxide

Dust storm:
up to 50 km high

Surface pressure:
1/150 of Earth

Water ice clouds:
10 to 25 km high

Meteors streak into Mars' thin carbon dioxide atmosphere (above). Winds near the surface can whip up planetwide dust storms. Yet air pressure on Mars is low— about as Earth's would be at a height four times that of Mount Everest. Craters scar Mars and its two moonlets (right).

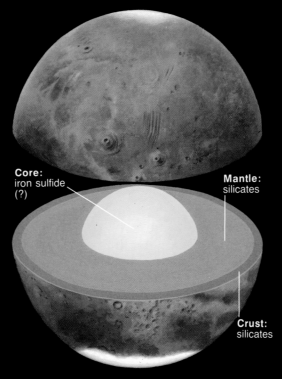

Core:
iron sulfide
(?)

Mantle:
silicates

Crust:
silicates

A rocky interior that's low on metal gives Mars only one-tenth Earth's mass, two-thirds its density, and weak surface gravity—just over a third that on Earth.

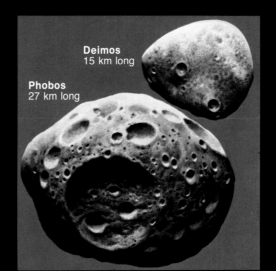

Deimos
15 km long

Phobos
27 km long

Ice caps of frozen carbon dioxide—dry ice—grow in Martian fall and shrink in spring (below). As one cap expands, the other retreats. Both poles also have permanent caps. At the south is a cap of frozen carbon dioxide plus water ice. The year-round northern cap is water ice.

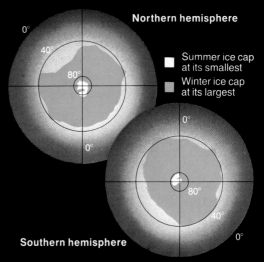

Northern hemisphere

0°
40°
80°

■ Summer ice cap at its smallest
■ Winter ice cap at its largest

0°

0°

80°

Southern hemisphere

40°

0°

Earth

Mars
Diameter:
6,794 km

More than any other planet, Mars has caught our imagination. One reason is the possibility of life on the Red Planet, now or sometime in its past.

We set out on our voyage to Mars when the Red Planet is closest to Earth, some 56 million kilometers away. Earth and Mars make side-by-side approaches about every 780 days, but the closest occurs only once every 15 or 17 years.

As we near the planet, we see two gleaming ice caps at the poles. Between them the globe is mottled in varying shades of red. By timing the motion of land features across the face of the planet, we confirm that Mars rotates once every 24 hours and 37 minutes. So a Martian day is nearly the length of ours. But Mars' slower orbital speed—and, more important, its larger orbit—give the Red Planet a longer period of revolution. A Martian year lasts 687 Earth days. Mars is only a little more than half the size of Earth, and its weaker gravity would cause a 100-pound Earthling to weigh only 38 pounds on Mars.

Martian secrets revealed

It takes us about 15 minutes to descend through the atmosphere and touch down on the surface. We find that most of the air is carbon dioxide. Nearly all of the Red Planet's water supply is frozen in the ground and locked up as ice in the polar caps. In fact, an ice age seems to grip Mars.

On our way down through the Martian atmosphere we saw countless craters marking the surface. They range up to several hundred kilometers across. We notice that they have been worn much smoother than the craters on our Moon. We sample the Martian soil and find much that is familiar to us in Earth rocks. In chemical composition the soil is mostly in the form of the elements silicon and oxygen bound up with metallic elements such as iron and magnesium. The grains of soil appear to be coated with iron-oxide rust. That hue, seen from Earth, long ago gave Mars the name Red Planet. In its southern hemisphere violent summer winds sweep up great clouds of reddish dust, which sometimes hide the entire Martian surface for months.

Volcanoes and canyons

In the equatorial region of Mars we climb up onto a broad dome that measures about 5,000 kilometers across, a thousand more than the distance between New York and Los Angeles. Called Tharsis, it rises about seven kilometers above the cratered surroundings. Four giant volcanoes soar up out of the Tharsis upland. Two lie north of the equator, one on it, and one to the south. At 27 kilometers their tops are three times higher than Mount Everest.

White clouds of water ice often shroud their slopes and can be seen from Earth by telescope. The largest volcano, Olympus Mons, is so huge that it would take us more than six hours to drive across its base at our highway speed limit of 55 miles an hour. This may be Mars' newest volcano, about 200 million years old. But the dating is uncertain; it may be much older. Many other ancient volcanoes are found here and elsewhere on Mars.

Along the eastern flank of the Tharsis dome is the series of canyons called Valles

Rock and barren dunes spread away from the Viking 1 lander, unlimbered for action on this Martian summer day— July 23 on Earth. Boom at right samples weather data; struts at left support an antenna. Cameras (housed in silos seen on page 129) send their wondrous imagery to the Viking orbiter when it comes into range, and the orbiter sends the images to Earth—a 20-minute trip across 360 million km at the speed of light. Weeks after this day, Viking 2 landed and sent data until 1980. Viking 1 communicated with Earth until 1982.

Marineris. In places the canyons are four times deeper than the Grand Canyon, and in length would span the United States. The width varies up to 500 kilometers. Most experts think the rifts were formed by massive marsquakes, then eroded by landslides, wind, and possibly water.

We find volcanoes in the southern hemisphere, but not as many as in the north. Some impact craters have been blasted into their slopes. This shows that the volcanoes were active early in the history of the planet, when meteoroid bombardment was still quite heavy.

The evidence of that bombardment is most noticeable in the southern hemisphere, heavily pocked with craters. Photographs sent by the first three Mariner spacecraft made it seem as if the whole surface of Mars had the same rugged look. But they had shown only a part of the planet. In 1972, after waiting for one of those giant dust storms to clear up, Mariner 9 photomapped the entire globe. We then learned that the Red Planet's surface divides into two quite different landscapes. If a line were drawn between them, it would cross the Martian equator at an angle of about 35 degrees. So the two landscapes are not quite the same as Mars' northern and southern hemispheres.

We see few craters in the northern landscape. This area has been heavily worn and

On the next page: *Wind-scoured Valles Marineris scar the planet's middle. The valleys, named for the Mariner flights, total 5,000 km in length, compared to 450 for Arizona's Grand Canyon system. A dust storm kicks up in this scene based on images from Mariner 9 and Viking.*

resurfaced by lava flows and sediments carried from other parts of the planet.

By contrast, parts of the southern landscape are cratered to saturation. The most heavily cratered areas appear to have received a bombardment equal to that on the bright highlands of our Moon. Rocks brought back from those highlands suggest that the blasting occurred more than four billion years ago. If Mars was bombarded at that time, as seems likely, then nearly half its landscape is very ancient. Some major landforms have remained largely unchanged for billions of years.

Right in the middle of the heavily cratered surface is a spectacular sight. Called Hellas, it is a sprawling circular basin that could swallow up Alaska with room to spare. A giant bowl up to four kilometers deep, the Hellas basin probably was excavated when a meteoroid several kilometers across smashed into Mars.

As the Martian seasons change, so do the polar ice caps. During spring in the northern hemisphere the ice cap shrinks and by midsummer nearly disappears; meanwhile the southern hemisphere ice cap grows larger with the autumn and winter seasons there. Most of the polar ice is frozen carbon dioxide—"dry ice" as we call it—with a little water ice. During spring in one hemisphere, the polar cap shrinks because increasing solar heat turns the carbon

131

What mighty fountains heaped up the lava dome of Olympus Mons, 20 times bulkier than any Earth volcano? In this painting ice clouds climb the slopes as the atmosphere cools. The caldera, rings within rings, spreads 80 km across. The base would blot out the heart of Texas.

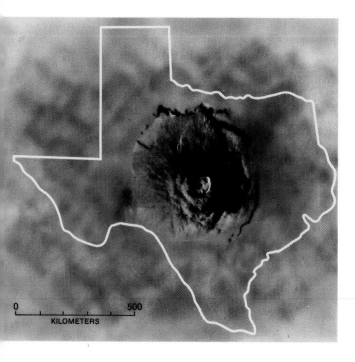

0   500
KILOMETERS

dioxide ice to gas. The gas enters the atmosphere. At the opposite, cooling pole the gas freezes and builds up a cap. The small patches of ice left at each pole during their summers differ from each other. In the north, the permanent patch is water ice. In the south, the cap is carbon dioxide ice, with some water ice. The variation in distance from the Sun explains the difference.

Potato moons

Mars has two moons, each of which looks like a giant orbiting potato. More than 300 years ago Johann Kepler guessed that Mars had two moons, though he had not a shred of evidence for his claim. Kepler figured that since Earth had one Moon and Jupiter had four (all that were known then), the planet in between—Mars—

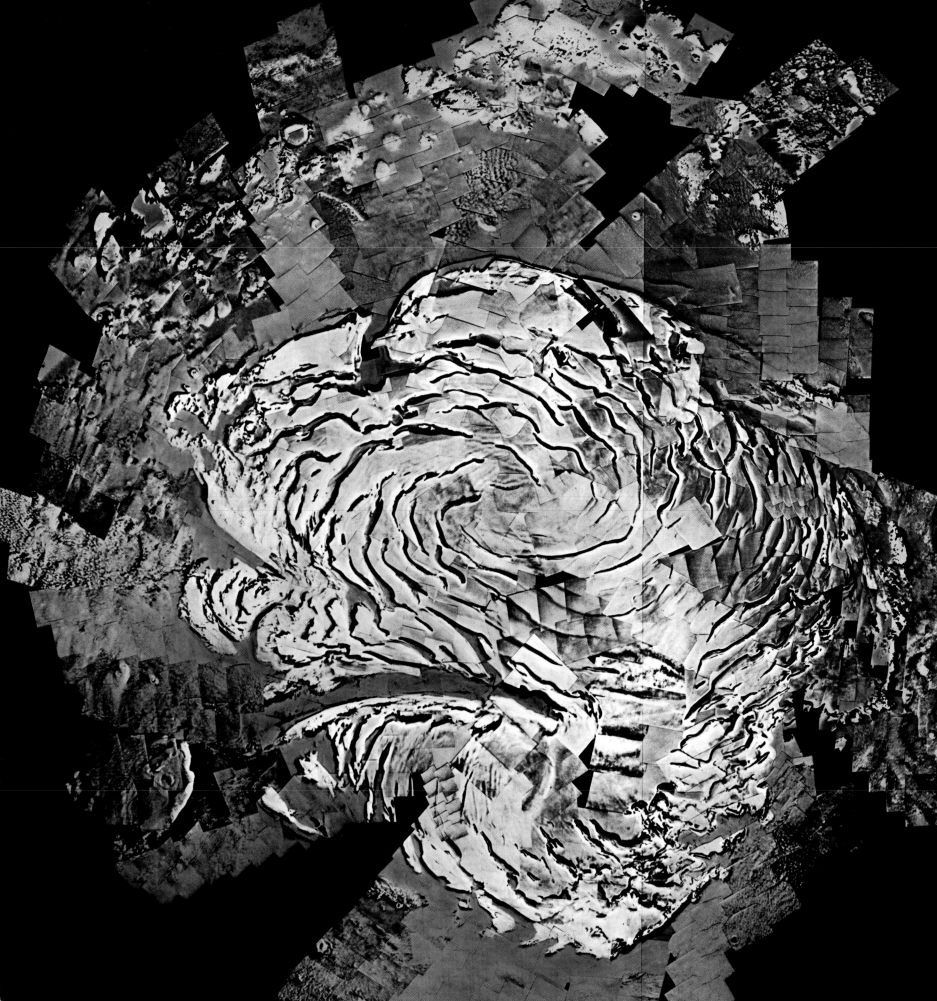

ought to have two. It wasn't until 1877 that the tiny bright dots were seen moving near Mars. The observer named them for sons of the Greek war god: Phobos and Deimos—Fear and Terror. Then, in late 1971, while a dust cloud made photographing Mars impossible, Mariner 9 put its waiting time to good use with a detailed scan of the satellites. At last it was possible to judge the real shape and size of the Martian moons.

Phobos speeds around Mars in 7 hours, 40 minutes. The more distant Deimos takes more than a day. From Mars' equator we see Phobos in the night sky for about 4 hours; Deimos takes 60 hours to cross the sky. From Mars the moons are seen to move in opposite directions. This is because Phobos takes less time to revolve once around the planet than Mars takes for a daily rotation, while Deimos takes a few hours more. From Mars' surface, Phobos appears one-third the size of Earth's full Moon, but tiny Deimos looks only like a very bright star.

Reading geological history

Geologists have tried to put together a history of Mars based partly on reports from unmanned spacecraft such as Viking and partly on what we know about Earth. According to one theory, the violence involved in the formation of the planets produced enough heat to partially melt the young planet. Heavy metal, mainly iron, sank to the center and formed a core. Silicate material formed a mantle. At the cool surface an early crust took a battering by meteoroids. During the first few billion years, decaying radioactive elements pro-

duced more heating. Melting of the mantle let lighter materials float upward and form a thicker crust.

This time of strong internal heating led to expansion of the globe, probably accounting for the fractures observed in Mars' crust—including the great Valles Marineris. This probably was also the time of maximum volcanic activity.

An outpouring of gases from the interior could have formed a denser atmosphere than we see today. Heated by the "greenhouse effect," such an atmosphere would contain lots of water vapor. Mars would have enjoyed a period, perhaps lasting millions of years, when rain fell on the surface.

This would have led to water erosion, including the formation of networks of drainage channels. When the ice-age conditions we see today returned, liquid water could no longer survive at the surface. It was lost to the permanent polar ice caps and to underground permafrost. Afterward, as a result of volcano activity, or perhaps the impact of large meteoroids, some of the permafrost was melted in places—causing sections of the surface to collapse and the release of lots of water. Raging floods then cut great channels into the crust.

What about life?

If Mars once had water, did it have life? Does it today? Living things on Earth must have liquid water to survive. On Mars' surface, liquid water does not exist. Mars' atmosphere is so dry that water vapor cannot reach the pressure needed to turn to liquid. So when ice is heated, it passes into vapor without melting. By one estimate, if all the

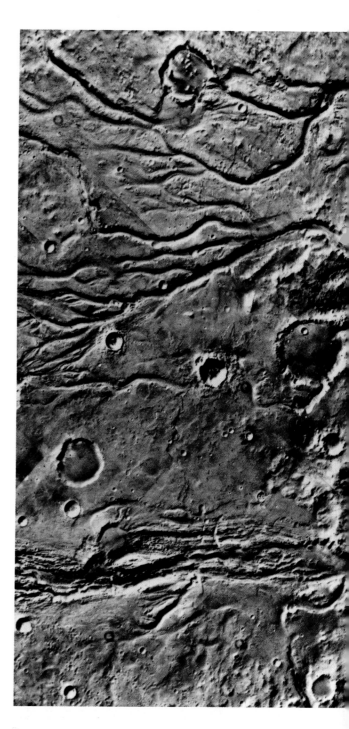

Mars

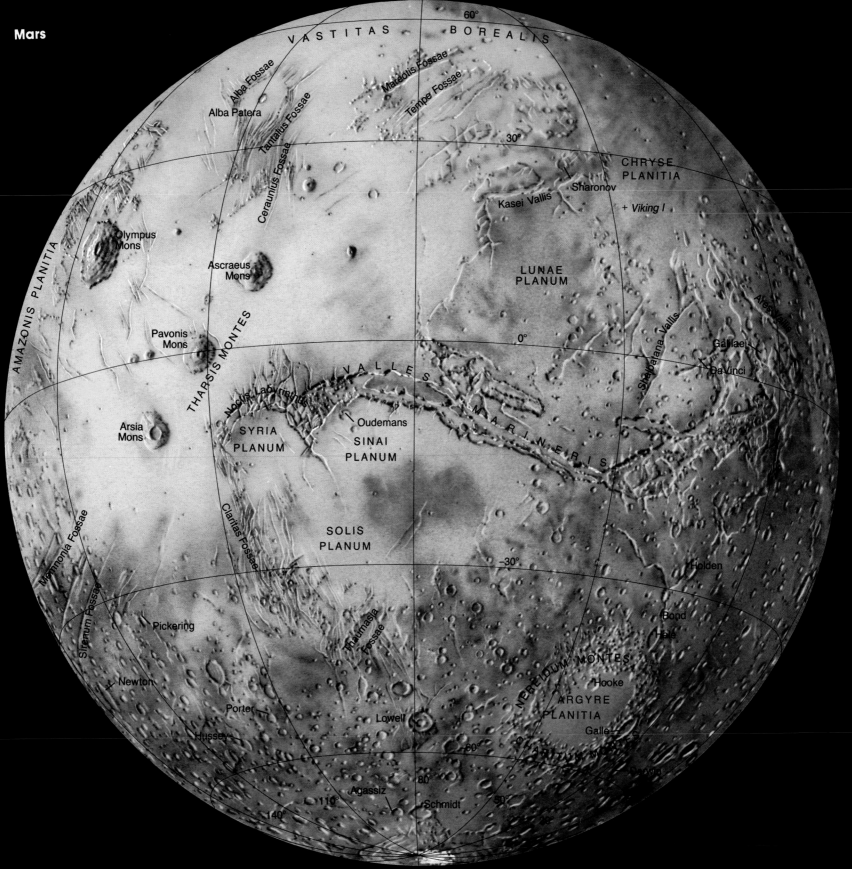

VASTITAS   BOREALIS
60°

Alba Fossae
Alba Patera
Tantalus Fossae
Ceraunius Fossae
Mareotis Fossae
Tempe Fossae
30°

CHRYSE
PLANITIA

Sharonov
Kasei Vallis
+ Viking I

AMAZONIS PLANITIA

Olympus
Mons

Ascraeus
Mons

LUNAE
PLANUM

Ares Vallis

THARSIS MONTES

Pavonis
Mons

0°

Galilaei

Shalbatana Vallis

Da Vinci

Noctis Labyrinthus

VALLES

Arsia
Mons

SYRIA
PLANUM

Oudemans

SINAI
PLANUM

MARINERIS

Memnonia Fossae

Claritas Fossae

SOLIS
PLANUM

-30°

Holden

Sirenum Fossae

Pickering

Thaumasia
Fossae

Bond

Hale

NEREIDUM MONTES

Hooke

Newton

ARGYRE
PLANITIA

Porter

Lowell

Galle

Hussey

60°

CHARITUM MO

80°

Agassiz

110°

Schmidt

50°

140°

138

The computer as artist paints an eerie afterglow to a Martian sunset. The Sun had sunk below the horizon; in the image sent by Viking 1's lander, the land lay dark. As the computer "stretched" data to detail the foreground, Mars flamed and colors rippled out from the ebbing light.

vapor in the atmosphere could turn to water, it would fill a small lake. The kind of life we know could not exist on Mars today. Maybe it did once. Maybe it had time to adjust to the new conditions, as the planet became dry, and remains in existence somewhere on Mars.

The Viking landers tried to find evidence of life in five different ways. One was with cameras. They sent numerous pictures, but none revealed "macrobes," the larger life forms, plant or animal, living or fossil, that we could recognize.

Another way was with an instrument to analyze soil, to detect organic molecules containing carbon. Such molecules are the building blocks of our living things. The Martian soil revealed none. Viking also carried three laboratories to test for tiny life forms, microbes, in the soil.

One test tried to find out if anything in the soil changed carbon dioxide into chemical compounds the way our plants do. In another little lab, water and nutrients were put into a soil sample to see if anything in it released oxygen or carbon dioxide the way plants or animals in our soil do. A third lab tested for anything in the soil that might consume chemical compounds the way humans and other animals do.

Although the experiments did not give us entirely clear answers, there is general agreement that Viking did not detect life. But the results from only two locations cannot rule out the possibility that Mars somewhere shelters life — or once did.

Perhaps this most fascinating of all Martian mysteries will not be solved until we send astronauts to explore the Red Planet.

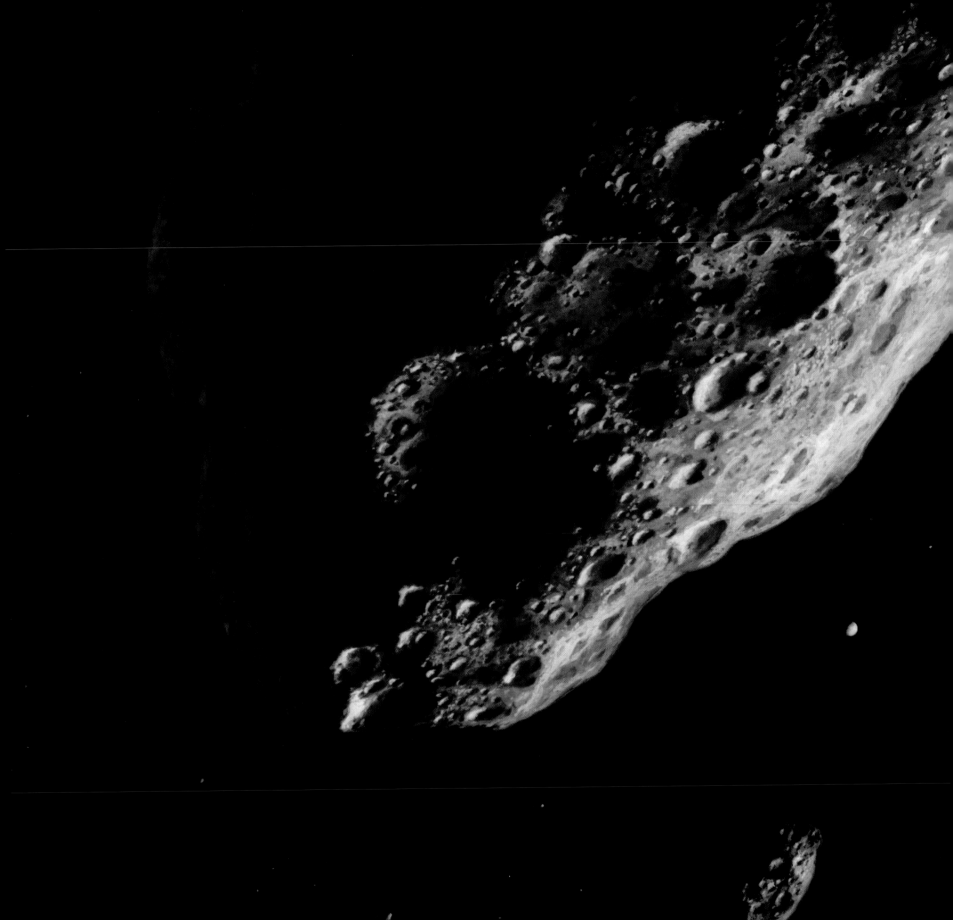

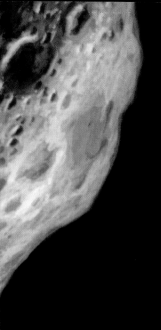

A Planet That Never Was

Asteroids & Meteors

Some call them "minor planets." Others have called them "planetoids." They are the asteroids, tens of thousands of tumbling, sometimes bumping lumps of rock-metal fragments that wheel around the Sun, most of them in a broad orbit between Mars and Jupiter. About 102 million kilometers beyond the orbit of Mars we come to the inner edge of this belt of asteroids. It stretches on for about 165 million kilometers toward Jupiter's orbit.

Here, as we look toward rosy, banded Jupiter, we can see that there is a lot of space between individual asteroids. Although we have never landed on an asteroid, we

have gone close to a few of them. They are pitted with many craters, like Mercury or our Moon.

During the early ages of our Solar System, thousands of these cosmic bombs plunged into Mercury, Venus, Earth, Mars, and Jupiter, as well as into the planets' moons and into each other. And most likely small pieces of them—the meteoroids—will keep doing so from time to time over the billions of years left in the life of our Solar System.

Because the asteroids batter each other, and because they are so small, few of them are neatly sphere-shaped.

Facts about
Asteroids

An asteroid, by tradition, is named by its discoverer. One honors England's Queen Victoria, another the mythical Icarus who flew too near the Sun.

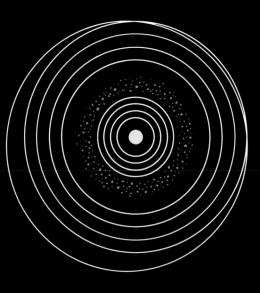

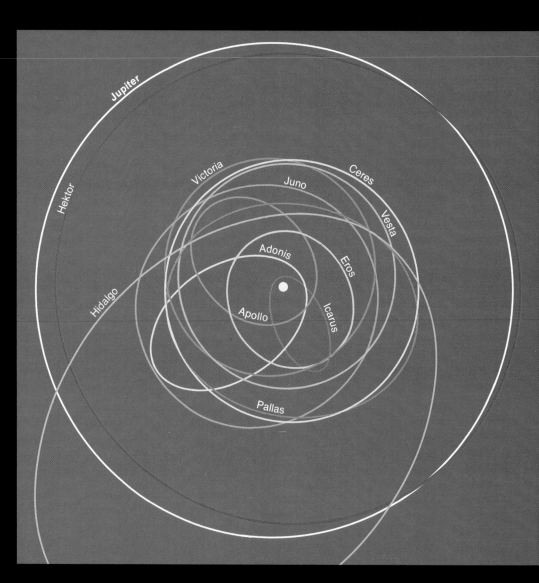

There are as many asteroid orbits as there are asteroids. Many of these objects, like Ceres, stay within the asteroid belt, a wide band between Mars and Jupiter. A planet ought to be there, astronomers once said; when they found asteroids instead, they thought these must be its pieces. But that's unlikely. All known asteroids lumped together would still be smaller than the Moon. Now and then a big one crosses Earth's orbit. Every million years or so, three or four of them ram into Earth and blast out craters many kilometers across.

144

Juno

Ceres

Earth's Moon

Davida

Phobos

Pallas

Hektor

Psyche

Eunomia

Laetitia

Vesta

Eros

Interior

Crust

Small asteroids have irregular shapes and the same composition outside and in—nickel and iron, or stone, or a little of both. Some asteroids may also contain large amounts of carbon, which gives them a dull, blackish color. Asteroids range in size from big Ceres to tiny lumps. On Ceres you would weigh 1/33 your Earth weight—and on a very small asteroid you would weigh so little that you might jump up and never come down.

Oddballs of the Solar System, asteroids come in many sizes and shapes. First to be discovered was Ceres, the largest known. Over 3,300 are now numbered; hundreds of thousands more may whirl about, too small to be seen. Phobos, a moon of Mars, may be a captured asteroid. Some asteroids are so massive their own gravitation has molded them into spheres. Collisions may shatter them or join them into odd shapes.

A small asteroid crashes into one 10 times its size. The larger asteroid will suffer only a few new craters but the smaller asteroid is shattered. In 1983, the Infrared Astronomical Satellite (IRAS) found dust bands around the asteroid belt, probably the result of collisions within the belt.

The asteroid Ida and its moon—the first such combination ever discovered—were photographed in 1993 by the Galileo spacecraft on its way to Jupiter. The gravitation of the tiny moon is so slight that a person could leap into orbit from its surface.

In the late 1700's German astronomers, calling themselves "celestial police," began to search with telescopes for a missing planet. They did this because in 1772 an astronomer, Johann Bode, had publicized a mathematical law, invented by Johann Daniel Titius, which gave the relative positions of the planets in the Solar System. However, something was amiss. In the gap between the orbits of Mars and Jupiter, Bode said there should be a planet.

Minor planets
On January 1, 1801, the Sicilian astronomer Giuseppi Piazzi found something which he thought was the missing planet. Later named Ceres, it turned out to be a mere pebble of a planet about 945 kilometers across—a third the size of the Moon. Then in 1802 another small planet, Pallas, was discovered in the gap. It was half the size of Ceres. In 1804 a third such object, Juno, was found—less than half the size of Pallas. Then in 1807 a German astronomer saw a fourth, Vesta—the brightest asteroid and the only one that can be seen with the naked eye. By 1890 astronomers had found 300 of these planets in the Mars-Jupiter gap. Today we know the orbits of more than 5,000. There may be hundreds of thousands. Most are boulder-size to mountain-size lumps of rock and metal that we call *asteroids,* or *minor planets.*

Earth-crossers
The average time an asteroid takes to go around its orbit is about five Earth years. A few have long, stretched-out orbits that take them very close to the Sun—and to

Earth. These are called Apollo Objects, or Earth-crossers, because they swoop in so close and intersect the orbit of Earth from time to time. (Their name comes from the asteroid Apollo, the first seen to cross our orbit.) In all, we know of about 40 Apollos. Some astronomers suggest that certain Earth-crossers may be the remains of old comets, while others are asteroids.

Eros hurtles through space, spinning rapidly at the rate of once every 5.27 hours. It has come within 23 million kilometers of Earth. Apollo has come within about four million kilometers of us. And Hermes has zoomed to within 770,000 kilometers, only twice the Moon's distance from Earth.

Astronomers don't worry too much when the Earth-crossers sweep in close to us. The chance of one of these cosmic mountains crashing into us is pretty small—a catastrophic hit perhaps every 65 to 100 million years. Some experts think the interval may be much shorter. But if a

giant should strike, it could produce an explosion as great as 20,000 megaton hydrogen bombs, and might blast out a crater hundreds of meters deep and as broad as a large city. Many scientists believe that just such an impact 65 million years ago led to the extinction of the dinosaurs.

We had one near miss on August 10, 1972, but this bright meteor, or bolide, was not a giant. Thousands of people saw a bright streak of light blaze a path in broad daylight through the sky (above) over the United States and Canada. It whizzed on out of our atmosphere again, but it had come as low as 60 kilometers and created

sonic booms. Based on factors such as speed and brightness, some scientists estimated that the object measured 10 meters across and weighed 1,000 tons.

Where did these rock and metal chunks the size of footballs, houses, and mountains come from originally? Astronomers used to think that early in the history of the Solar System there was a planet in the gap between Mars and Jupiter. Gradually the planet was pulled so close to Jupiter that Jupiter's powerful gravitation shattered the smaller planet, breaking it up into the asteroids. Now they think that the asteroids may be chunks of rock and metal left

over from the time the planets were being formed. Jupiter's gravitation, they say, prevented the asteroid fragments from ever collecting into a planet-size object.

Messengers from space

On the night of November 13, 1833, a shower of meteors lighted up the sky. "The stars fell like flakes of snow," said one observer. Others, also believing that meteors were falling stars, thought that no stars would be left in the sky the next night.

Away from the city lights on any clear night you can see about five "falling stars" in an hour. They are not stars, of course, but lumps of rock and metal which range from the size of dust grains to—once every several thousand years—a rare piece more than 10 meters across. When they are in space these objects are called *meteoroids*. Most of them probably come from the asteroid belt. When one enters Earth's atmosphere at a speed from 15 to 72 kilometers a second it burns up from frictional heating and glows as the quick streak of light that we call a *meteor*. If it survives the hot journey through the atmosphere and strikes the ground, we call it a *meteorite*.

Every day meteorites plunge to Earth and add at least a quarter of a ton to our planet's mass. The largest meteorite found on Earth is in Namibia. It weighs some 60 tons. The next largest, about 30 tons, is on display at the American Museum of Natural History in New York. Another 10 tons a day are added by space-dust particles called *micrometeorites,* which are captured by Earth's magnetic field and float gently to the ground. Some of these have even been found on the petals of flowers. If you are lucky, someday you may find a micrometeorite on a magnolia.

Sporadic meteors are the ones we see on just about any night, falling out of the sky from any direction. But others, the *shower* meteors, seem to travel in swarms. The swarms return year after year and shower down on us at the rate of about 60 an hour from certain parts of the sky. Those showers that seem to come out of the constellation Leo are called the Leonids; those coming from the direction of Orion are called the Orionids; and those we see in the constellation Gemini are called Geminids.

At 2,758 kg, the Old Woman (above) is the second largest meteorite ever found in the U.S. Made of iron and nickel, it was discovered by gold prospectors in 1976. Marines pry it out of the Old Woman Mountains in California, where it fell hundreds of years ago.

This fragment of the stony-iron Mt. Padbury meteorite was photographed through a microscope in polarized light. The microscope blocks out certain types of light, giving the bright colors. Found in Australia, the Mt. Padbury meteorite may be a piece of the asteroid Vesta.

149

Perhaps 50,000 years ago a meteorite the size of a railroad car crashed into a desert, producing a crater 1.3 km wide. It blew out about 400 million tons of rock. Meteorites this large rarely collide with Earth. Barringer Crater, Arizona (left), resulted from this impact. Earth's surface has about 80 large meteorite craters. Most of them lie in highly populated areas, where people are likely to find them. Undoubtedly the jungles hide many more. Scientists have determined that several unusual meteorites recently found in Antarctica came from the Moon.

Shower meteors seem to radiate toward us from a point in a particular constellation, but this is an illusion. Actually, they travel in parallel lines, just as snowflakes fall in more or less parallel lines but appear to radiate from a particular point as we see them through the windshield of a moving car. Also, we see more snowflakes ahead of us, through the windshield, than when we look through the rear window. Likewise, we see more meteors when we look in the direction of Earth's orbital motion—Earth's "windshield"—than when we look "back." This happens, in our daily rotation, between midnight and noon, so the best times for meteor watching are the dark hours between midnight and dawn.

The role of comets
Astronomers think that most shower meteors are debris from comets. The Leonid and Perseid showers of late summer are probably the remains of comet tails. And the Orionid shower may be dust left from past appearances of Halley's Comet. The Infrared Astronomical Satellite (IRAS) added a new element to this idea when it found an asteroid whose orbit crosses planet Mercury's. The asteroid's orbit looks like that of a comet, and scientists soon discovered that the asteroid's path coincides exactly with the orbit followed by the Geminid shower meteors. Does that mean the Geminid shower is due to an asteroid instead of to a comet? Or does it mean that the new "asteroid" is actually a burned-out comet, as certain Earth-crossers might be? If so, the Geminids also come from a comet.

Classes of meteorites
If meteoroids and asteroids really did form at the same time as the planets, we would expect to find them made of some of the same materials as the planets. And this is just what we do find. There are three classes of meteorites—iron, stony, and the stony-irons. The irons are a blend of iron and nickel. The stony meteorites are mostly silicate rock with a little iron and nickel. Midway between the irons and stony meteorites are the stony-irons, made up of about half stone and half metal.

Most meteorites found on Earth are stony, as are about 90 percent of those that fall through space. But the stony meteorites, once on the ground, are harder to recognize than the irons. They also decay faster—both during their passage through the atmosphere and because erosion attacks them once they land. The irons are tougher but, since they are so outnumbered, we find fewer of them.

Off to the asteroids
Scientists still have many questions about what asteroids are and where they come from. NASA hopes to send several spacecraft past asteroids to gather information at close range. Galileo, on its way to Jupiter, captured the first good look at an asteroid—Gaspra. A small body about the size of the Mars satellite Deimos, Gaspra is a fragment from a 200-million-year-old cosmic collision. On the Near Earth Asteroid Rendezvous mission, a spacecraft will fly by the asteroid Iliya and, in 1999, will follow the asteroid Eros to study its surface, structure, and composition.

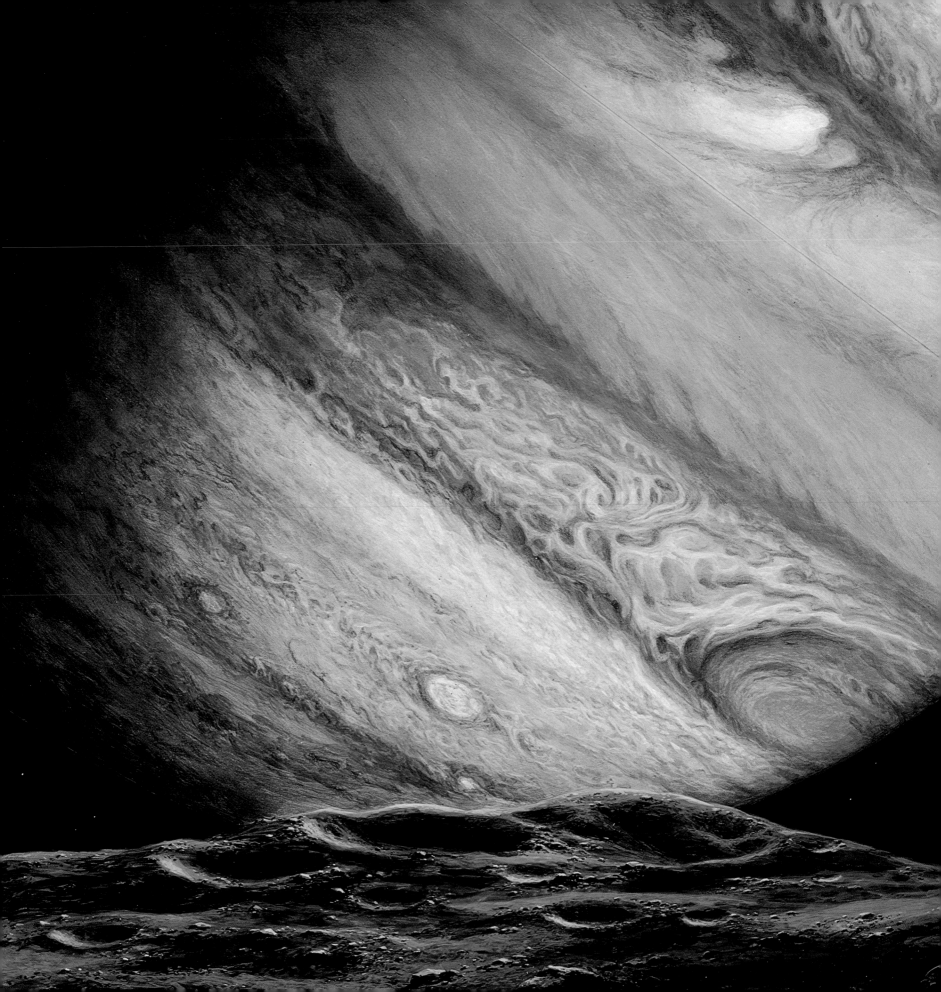

King of the Planets

Jupiter

Jupiter's colorfully banded atmosphere is ever churning, ever changing. This would be our view of the colossal, belted globe from Amalthea, one of its inner moons. Amalthea is a reddish, rocky little world about 110,000 kilometers from the top of Jupiter's clouds.

Jupiter's most prominent feature is its Great Red Spot, a hurricane-like storm that has probably been raging for centuries. At lower left are three smaller, white storm areas. The soft colors of Jupiter's bands shimmer against the black sky. These bands may be upwelling atmospheric gases composed of giant molecules.

This king of the planets is one and a half times larger than all the others put together. It is also the fastest spinning planet.

Pioneer and Voyager flybys of Jupiter told us much about the planet. The two Voyagers discovered its 14th, 15th, and 16th moons, but they found fewer moons in Jupiter's orbit than in Saturn's. The Voyagers also discovered Jupiter's ring system. (Here we see it as a thin line slanting along the equator.) But they left many interesting questions: How did Jupiter get its rings? Does Jupiter's atmosphere support primitive life forms? What is the planet like beneath its massive cloud cover?

Facts about
Jupiter

King of Roman gods: Jupiter. His name is a fitting one for our largest planet. A traditional sign for his lightning bolt gives the planet its symbol, ♃

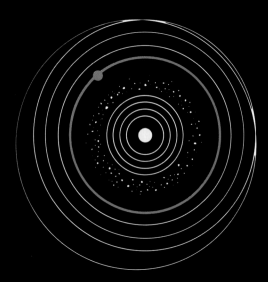

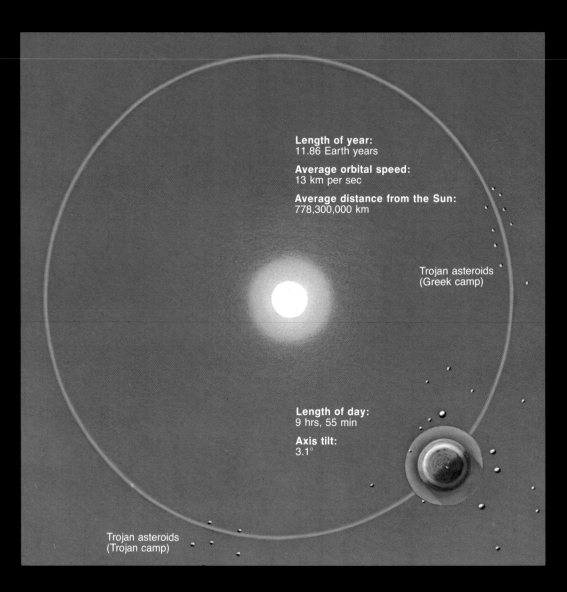

Length of year:
11.86 Earth years

Average orbital speed:
13 km per sec

Average distance from the Sun:
778,300,000 km

Trojan asteroids
(Greek camp)

Length of day:
9 hrs, 55 min

Axis tilt:
3.1°

Trojan asteroids
(Trojan camp)

Master of its orbit, the fifth planet holds a faint ring system and at least 16 moons in its gravitational grip. Four moons are as big as small planets. The combined pulls of Jupiter and the Sun also keep two asteroid groups, called the Trojans, in Jupiter's orbit. One group moves along a sixth of the way ahead of Jupiter, the other equally far behind. They bear heroes' names from the ancient Trojan War—the first group represents Greece, the second Troy. Jupiter's tiny outer moons may be Trojan asteroids, trapped when they strayed too close to the planet.

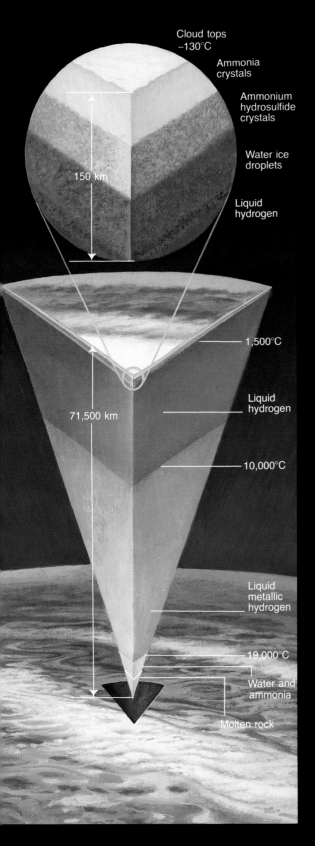

Cloud tops
−130°C

Ammonia
crystals

Ammonium
hydrosulfide
crystals

Water ice
droplets

Liquid
hydrogen

150 km

1,500°C

Liquid
hydrogen

71,500 km

10,000°C

Liquid
metallic
hydrogen

19,000°C

Water and
ammonia

Molten rock

Heavyweight champion of the worlds,
Jupiter (right) accounts for more than
two-thirds of all material in the Solar
System outside the Sun. It would take
318 Earths to equal Jupiter's huge mass.
Gravity two and a half times stronger
than our own creates intense pressures in
the swirling gases of its atmosphere.
The Voyagers found one bright ring
flanked by a faint vertical-extending ring
and an even fainter "gossamer" one.

An icy frosting of clouds (left), several
layers thick, covers what most of Jupiter
is made of—hot liquid hydrogen. The
atmosphere above is cold, but farther
down it could be well over 1,000°C.
In this pressure-cooker level, gases would
turn liquid and form a kind of steamy
slush—the top of an enormous hydrogen
ocean. About halfway down to the
planet's core, heat and pressure force the
hydrogen to act like molten metal.
Scientists think electrical currents in this
zone may create Jupiter's giant magnetic
field, which extends past the orbits of
several of its moons.

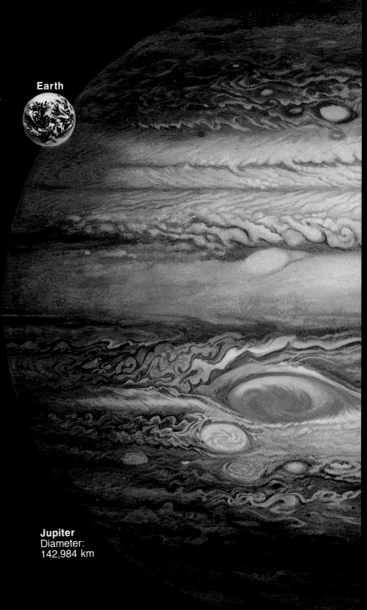

Earth

Jupiter
Diameter:
142,984 km

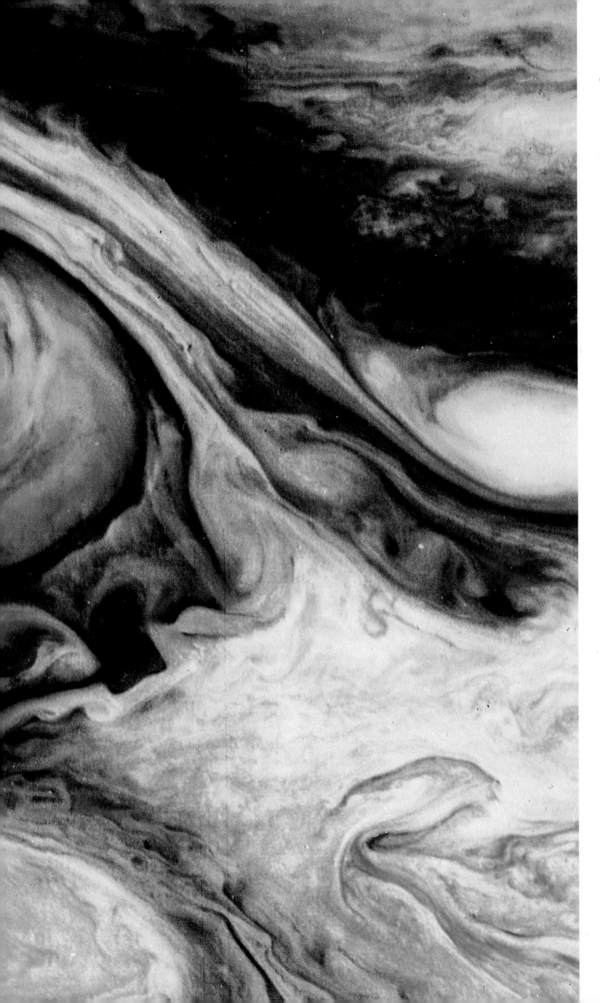

"The eye of Jupiter," astronomers call the Great Red Spot. The giant storm, its winds whirling up to 360 km an hour, appears in this Voyager view with some colors brightened to show details. To see how big the spot really is, set a grapefruit on the picture; that's Earth to scale.

On our make-believe journey through the Solar System we pass safely through the asteroid belt, just as the two Voyager spacecraft did in 1978. First they were flung out from Earth at a send-off speed of 52,000 kilometers an hour. Earth's and the Sun's gravitation gradually slowed them down. Eventually the Voyagers entered the gravitational field of Jupiter which sped them up. The Voyagers flew for one and a half years to reach Jupiter.

Jupiter is the largest planet in the Solar System. It is about five times farther from the Sun than Earth is. More than 11 Earths could be lined up along the giant planet's diameter and 1,300 Earths could easily be packed inside.

A giant planet with giant surprises

Seven million kilometers out from Jupiter, Voyager 1 entered the planet's powerful magnetic field. (See pages 158 and 159.) Voyager safely crossed the boundary where the solar wind clashes with Jupiter's magnetic field—an electrically turbulent area called the *bow shock*.

Jupiter's inner moons orbit within the magnetic field and are continually showered with high-speed charged particles, mostly protons and electrons. The current is spread too thin to harm a spacecraft, except perhaps in the highly-charged region between Jupiter and its moon Io. The movement of Io through the magnetic field sets up an electrical current five million amperes strong. The current flowing in a 100-watt light bulb is only one ampere.

The electricity linking Io with Jupiter is gradually eroding Io's surface. Gases

157

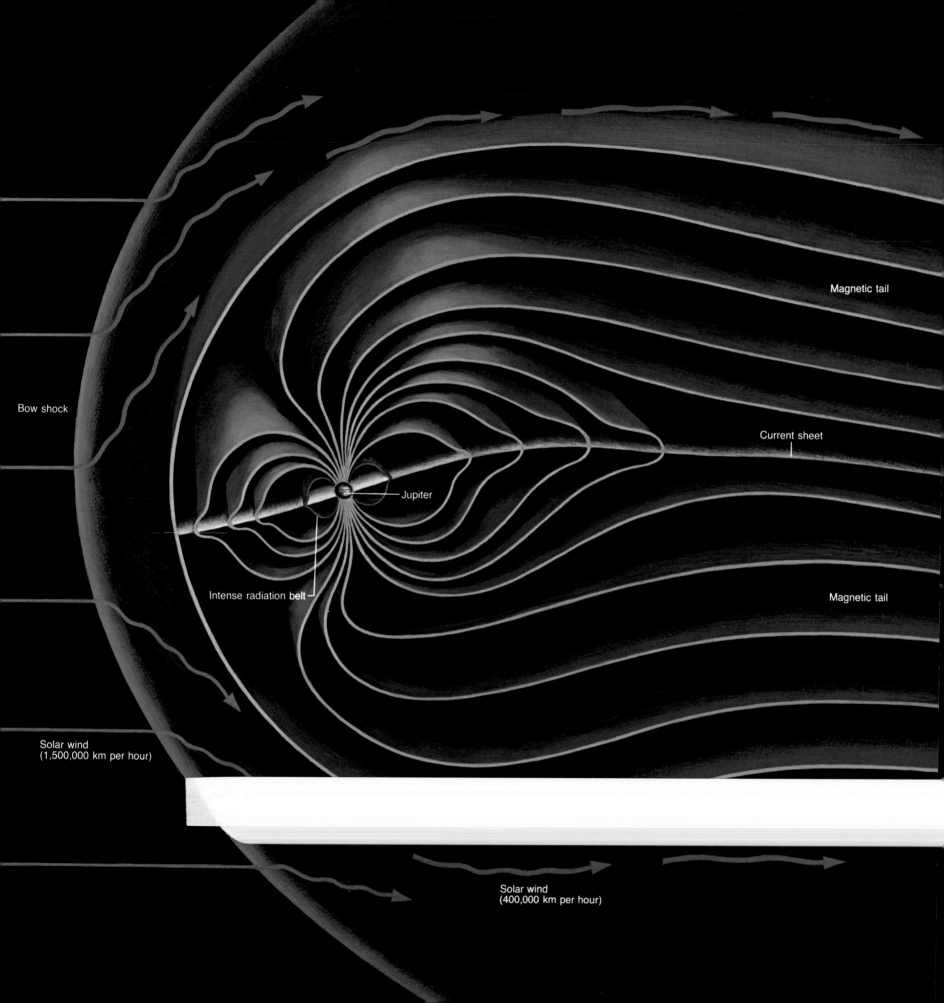

Magnetic tail

Current sheet

Bow shock

Jupiter

Intense radiation belt

Magnetic tail

Solar wind
(1,500,000 km per hour)

Solar wind
(400,000 km per hour)

Jupiter's invisible magnetosphere is far bigger than the Sun itself. If we could see it pulsing in our night sky, it would look about twice as big as the full Moon. The bow shock, where the magnetic field and solar wind meet, surges back and forth with gusts in the solar wind. A magnetic tail stretches outward — beyond the orbit of Saturn. Particles spun off from Jupiter's ionosphere collect in a thin sheet of electrical current (yellow). Near Jupiter, deadly radiation — thousands of times stronger than Earth's Van Allen belts — batters Jupiter's major satellites.

In 1979 NASA scientists watched in relief as Voyager 1 survived the magnetosphere's searing radiation and sent back floods of information about Jupiter. Voyagers 1 and 2 also began our exploration of the four big moons Galileo discovered: Io, Europa, Ganymede, Callisto.

explode out of Io's volcanoes and break up into charged atomic particles. Jupiter captures them, along with surface particles. They form an invisible doughnut-shaped electrical cloud around the planet. As Io circles Jupiter, the planet keeps the satellite a prisoner within this enormous energetic cloud.

A major surprise from the Voyagers was that Jupiter has a ring system. The faint main ring stretches from about 52,000 to 58,000 kilometers above the planet's cloud tops. An inner — halo — ring is even fainter, though thicker and more diffuse. The third ring, a tenuous, "gossamer" one, extends outward from about 58,000 kilometers above the cloud tops to almost the orbit of Jupiter's moon, Amalthea. What are the rings made of? Most scientists believe that they are composed of rocky debris from Jupiter's innermost moons, Metis and Adrastea, which are slowly being broken up by the planet's gravitation.

The Great Red Spot

Jupiter's atmosphere is a churning sea of rising and sinking clouds of many colors. Jet winds of super-hurricane force tear at the clouds and add to their motion. It has been compared to an enormous boiling kettle of brightly colored dyes that cannot be made to blend. The reason may be that these cloud cells contain different chemicals which originate at various depths within the lower Jovian atmosphere.

For more than 100 years astronomers have studied the Great Red Spot, a giant whirlpool of gas in Jupiter's southern hemisphere near the equator. They now believe it is a kind of weather phenomenon unlike any on Earth. Jupiter's turbulent winds drive the gas. And the planet's rapid rotation keeps the 40,000-kilometer-wide blob in a football shape. A prominent feature of Jupiter's atmosphere for at least 300 years, the Red Spot is large enough to swallow two to three Earths. Its color varies, making it more or less easy to see, but no one knows what makes it red.

The Great Red Spot revolves along with the rest of the cloud bands, but at a slower speed. Other cloud features sometimes catch up to the Red Spot and pass beneath it or around it. The spot also spins counterclockwise on its own axis, at the rate of one spin every six Earth days. So if you were watching Jupiter from one of its outer moons, you could almost use the Great Red Spot as a clock with hands of clouds that dissolve and form anew.

Below the Great Red Spot are three white oval storm areas. Astronomers watched them form only 40 years ago. Like the Great Red Spot, they are pretty much a mystery. There are many more of these storms. In 1973 the space probe Pioneer 10 photographed one in the northern hemisphere, the Little Red Spot. But a year later, when Pioneer 11 flew by, the Little Red Spot had disappeared. Voyager 1 saw its reappearance or, perhaps, a new spot.

The atmosphere

Because of its strong gravity Jupiter has kept much of its original atmosphere. The bulk of Jupiter's air is hydrogen, with helium making up about 10 percent. There are other gases as well, gases that may also

On the next page: *In this artist's view across the Great Red Spot, a chilly Sun and crescent Io float above a cloudscape vaster than any Earthly ocean. Below, lightning stabs into layers churning with chemicals needed for life. But life seems unlikely to form amid such violence.*

Jupiter

North North Temperate Belt

North Temperate Zone

North Temperate Belt

North Tropical Zone

North Equatorial Belt

Equatorial Zone

Equatorial Band

Equatorial Zone

South Equatorial Belt

South Tropical Zone

South Temperate Belt

South Temperate Zone

South South Temperate Belt

South South Temperate Zone

South Polar Region

GREAT RED SPOT

30°

0°

−30°

−60°

90°

120°

60°

150°

30°

180°

0°

Jupiter's swift rotation creates bands of cloud called belts and zones. Rivers of jet-stream winds rushing eastward carry the high-flying, light-colored zones. At lower altitudes, between the zones, dark-colored belts border jet streams that tear around Jupiter to the west.

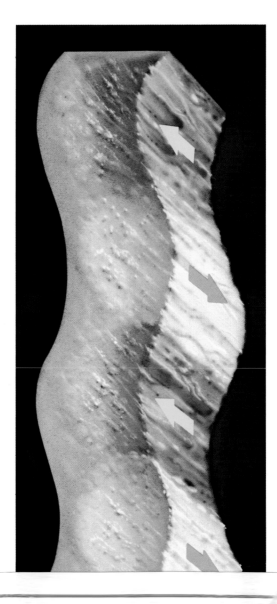

have been in the original atmospheres of Mars, Earth, and Venus. Those gases include methane, water vapor, and ammonia. The upper regions of Jupiter's clouds are cold, about –130°C, and the temperature of the Great Red Spot is even lower.

Into the clouds

All we can see of Jupiter is the top of a cloud deck about 100 kilometers deep. But the bottom of its atmosphere may be at a depth of 1,000 kilometers.

Jupiter probably lacks a solid, Earthlike surface. An explorer of Jupiter would first pass through a dense atmosphere of gaseous hydrogen that gradually changes to a strange lifeless ocean of liquid hydrogen. About 20,000 kilometers beneath this liquid layer, there may be a layer of hydrogen so dense that it acts like a metal. Beneath this layer Jupiter may have an Earth-size inner ball of rock and ice. Temperatures in this core region may reach 25,000°C, so the rock and ice are probably in a liquid state. The ice may be composed of water, ammonia, and methane; the rock of molten silicates. Pressures here may reach 40 million times the pressure at Earth's surface.

If this is what Jupiter's interior actually is like, then its metallic hydrogen layer probably carries the electric currents that produce the powerful magnetic field measured by both of the Voyager spacecraft.

Life on Jupiter?

As Voyager gradually moved into Jupiter's shadow zone it revealed more surprises. Toward the polar region of the planet grand auroras brightened the sky—the first auroras seen on any planet other than Earth. While Earth's auroras are caused by charged particles of solar wind, Jupiter's auroras seem to be caused by charged particles from its moon Io.

Voyager found that Jupiter's upper atmosphere is alive with lightning superbolts. Astronomers believe that this lightning may cause Jupiter's whistlers—bursts of radio noise.

Lightning bolts may also provide the atmospheric energy which triggers many of Jupiter's chemical reactions. Not far below the frigid cloud tops there must be a region that is "comfortably" warm—a region where water, methane, and ammonia gases react chemically, energized by the lightning. When these substances join they can form organic molecules, the chemical beginnings of life. Is it possible that within this zone simple living organisms evolved long ago and have adapted to a floating existence within the clouds? Most scientists are not hopeful. They think that Jupiter's rapidly churning air has prevented the development of atmospheric life forms, since the complex molecules necessary for life would be swept down into hot regions of the clouds and be destroyed.

If Jupiter's organic molecules have not combined and provided Jupiter with living organisms, they may enrich the planet in a less dramatic way. Rust-colored clouds well up and help color the Great Red Spot and redden some of the cloud bands.

Jupiter—a star that failed?

Jupiter emits more energy than it receives from the Sun—twice as much, Pioneer and

A Voyager camera captures two moons as they drift across a looming Jupiter. Sulfur colors Io (left), innermost of the Galilean satellites. Europa (right) displays an icy crust. Like our Moon, all four keep the same side inward, frozen that way in Jupiter's gravitational grip.

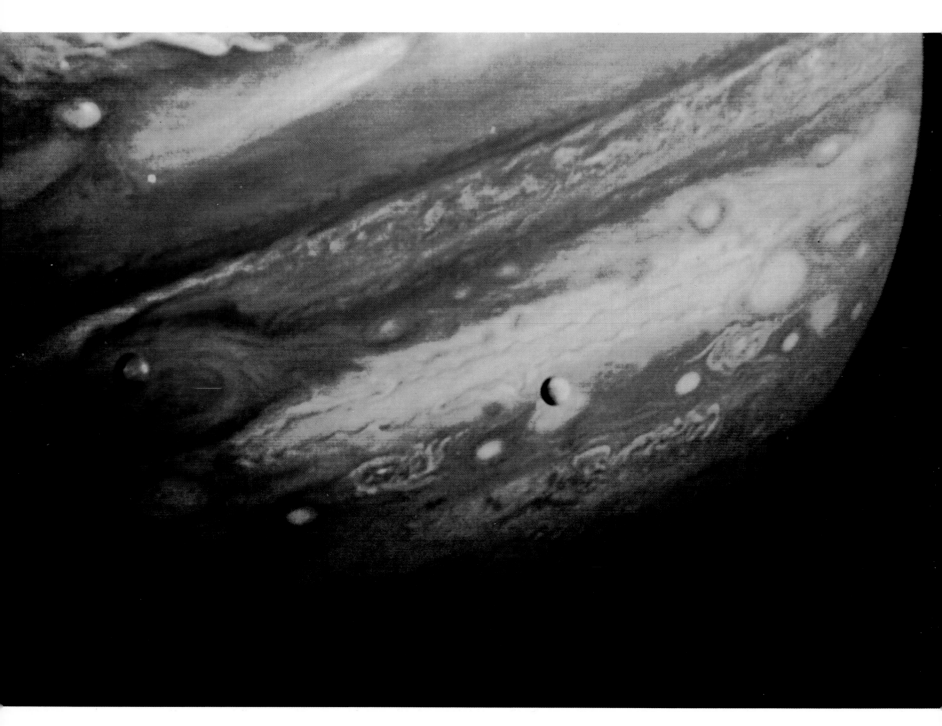

Jupiter's moons weave a jumble of orbits. An outer group (shown in blues) revolves retrograde, in tilted, elliptical paths that take as long as two Earth years to travel. The middle group (red colors) revolves direct—along with Jupiter's spin—in different tilts from the outer. These small moons may all be asteroids captured by Jupiter's gravitational field. The big Galileans (greens) follow more circular tracks, above Jupiter's equator. So do the little inner moons (yellows). Two—Metis and Adrastea—found by Voyager zip around Jupiter every seven hours.

Voyager found. Could Jupiter have been much hotter in its early history, maybe hot enough to warm its four large moons as the Sun warms the inner planets?

It seems so. Imagine the scene 4.6 billion years ago when the Solar System was taking shape. Jupiter, two and a half times as massive as all the other planets combined, began as an enormous gas ball that contracted and heated up, just as the infant Sun was doing. But, unlike the Sun, Jupiter had far too little mass to send its core temperature high enough to start fusion reactions. Instead of reaching the millions of degrees needed, the core heated up only a few tens of thousands of degrees. So Jupiter became only hot enough to glow cherry-red, like a red dwarf star, and for a while it bathed its inner moons in light and heat that faded as Jupiter slowly cooled.

Probably only the inner moons formed as satellites at the same time Jupiter formed as a planet. The eight small outer moons are all believed to be former asteroids captured by Jupiter's gravitation. And there may be even more moons in Jupiter's grip. Astronomers who are studying the data sent back by the Voyager spacecraft identified three new, small moons—Metis, Adrastea, and Thebe. They and Amalthea are the closest moons to Jupiter.

giant planet's four largest moons—Io, Europa, Ganymede, and Callisto, the moons discovered by Galileo on January 7, 1610.

Looking through his telescope, Galileo at first saw only three moons in a straight

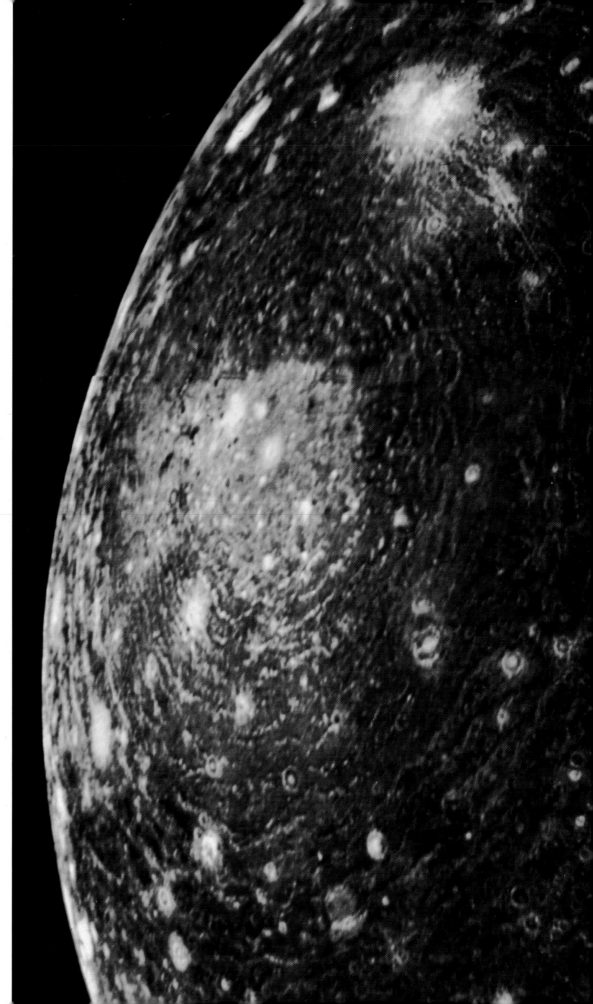

Ganymede (below)—Ice flung from fresh impact craters whitens Ganymede's surface. Dirt and rock darken older zones. Fault lines show that the thick ice crust may have broken into plates, like those of Earth. Ganymede is the largest Galilean moon, and the Solar System's largest.

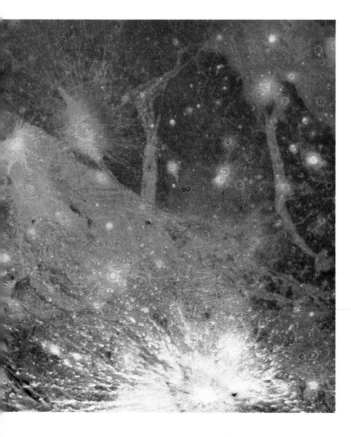

line across Jupiter, two at the left and one at the right. The next night he saw all three to the right. Galileo thought he was looking at distant stars and wondered how Jupiter had managed to move to the left of them. Two nights later he saw only two, and they were to the left of Jupiter. (The third was behind the planet.) Eventually Galileo counted four objects and realized that they were orbiting Jupiter: Their behavior made them moons, not stars.

The discovery caused great excitement. When Johann Kepler heard of Galileo's discovery he longed to see the new moons but did not have a telescope, and it was not

Callisto (left)—Like a frozen explosion, the huge Valhalla basin ripples Callisto's ice crust, which melted and refroze here after a giant meteoroid strike. Undisturbed by faults or volcanoes, the ancient face of the outermost Galilean moon may be the most cratered place in the Solar System.

Europa (below)—Since few craters mark the smallest Galilean moon, scientists think its cracked cue-ball landscape is relatively young. Water seeping through faults may have smoothed the surface ice, leaving the long stains. A global ocean may lie in the darkness underground.

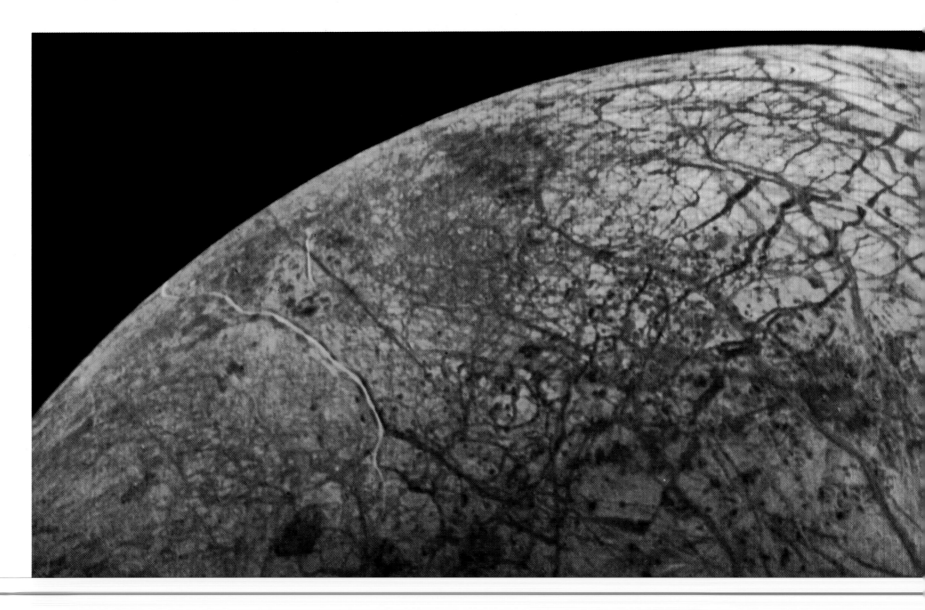

Io—Eroded by radiation, jolted by arcs of electricity, squeezed and stretched by tides . . . Io's surface reflects the moon's tumultuous state. The purple-plumed volcano on the horizon, called Loki, and others like it spew out enough sulfurous material to resurface the entire moon every million years. Sulfur compounds of various temperatures color Io's plains and calderas, accounting for the many hues apparent in this false-color image: Hot volcanic vents on the surface appear purple; volcanic flows, reddish orange; and cold, frosty plains, light blue.

until August 1610 that he finally did see them, and it was through a telescope that Galileo had made. Some astronomers refused to believe Galileo's claim. They felt that the Solar System with its seven "planets" was complete and were very upset that an age-old belief could be wrong. Some of them refused even to look through Galileo's telescope. But eventually even Galileo's bitterest enemies had to admit that Jupiter had moons.

Ganymede's grooves

Measuring 5,276 kilometers in diameter, Ganymede is Jupiter's—and the Solar System's—biggest moon. It circles the planet at a distance of about a million kilometers. Voyager 1 sent back pictures of the side of Ganymede that is locked to Jupiter. Voyager 2 photographed the opposite side. Both showed a satellite peppered with craters which make Ganymede look remarkably like our Moon. Many of the craters have bright rays. There is also a lighter terrain, with long grooves that look as though they were made by a giant rake pulled in curved and crisscross patterns. The parallel ridges and troughs of the grooved terrain are from 5 to 15 kilometers wide and run on for hundreds of kilometers. Scientists who try to determine the age of Ganymede's surface find that some areas are heavily cratered and lack grooves, while others are grooved but have few craters. It seems that the grooved terrain was present early in Ganymede's history and developed as meteoroid fragments were bombarding Ganymede.

One Ganymede feature, so big we can see it through telescopes from Earth, is a huge, circular area of old craters. Crossing the region are bright streaks, parallel and slightly curved. It is tempting to suppose that a meteoroid long ago blasted out these rings of ridges but that erosion erased most of them. There are also many "ghost" craters, mere stains of craters once prominent when Ganymede's icy crust was too warm and soft to preserve the original features. But near the south pole are large craters with sharp, fresh features. These must have been made sometime after the crust had cooled.

There are ice fields and great tile-like blocks that may be plates of ice. They range from several hundred to about a thousand kilometers across. Beneath Ganymede's crust, there may be a liquid water mantle and below that a solid core rich in silicate rock, similar to the rock making up Earth's crust.

Callisto's ringed terrain

Nearly twice Ganymede's distance from Jupiter is Callisto, 4,820 kilometers in diameter. The biggest attraction on airless Callisto is Valhalla, a gigantic group of rings raised by shock waves. A large meteoroid smashed into Callisto when it was still a young moon. The impact left a hole half as wide as Florida and perhaps 20 kilometers deep. When the monstrous chunk of rock and metal struck, it instantly must have melted a large area of Callisto's icy crust, sending out clouds of steam and creating a small sea. But the water would have frozen soon and left Valhalla as the shallow icy basin we see today.

Europa's icy plates

Europa is 3,126 kilometers in diameter, nearly as large as Earth's Moon. It orbits Jupiter at almost twice the distance that our Moon orbits us.

Europa appears to have an unusually smooth surface and only a few large impact areas. It is laced with dark, crisscrossing lanes tens of kilometers wide and several thousand kilometers long. What caused this moon-wide network? One guess is that Europa is still hot and active in its core. If it is, forces within the satellite would keep the crustal plates of ice moving about. Another guess is that Jupiter's strong gravitational grip—plus gravitational forces of the other satellites—keeps breaking Europa's icy plates and disturbing the material beneath. This may cause gases and dirt to float up to the surface where they freeze as the dark lanes of ice. Under them, about 100 kilometers down, there may be an ocean of liquid water, and beneath that, an active rock-metal core.

Volcanoes on Io

Io is one of the most fascinating objects in the Solar System. With a diameter of 3,632 kilometers, it is only slightly larger than our own quiet Moon. But Io is anything but quiet. The Voyager spacecraft clearly showed 10 violent volcanoes. There are many dead volcanoes on the Moon, Mars, Mercury, and Venus, but in the Solar System today only Io has more volcanic activity than Earth does. What makes Io so active? Astronomers are still studying Voyager data to find out more about this highly energetic moon.

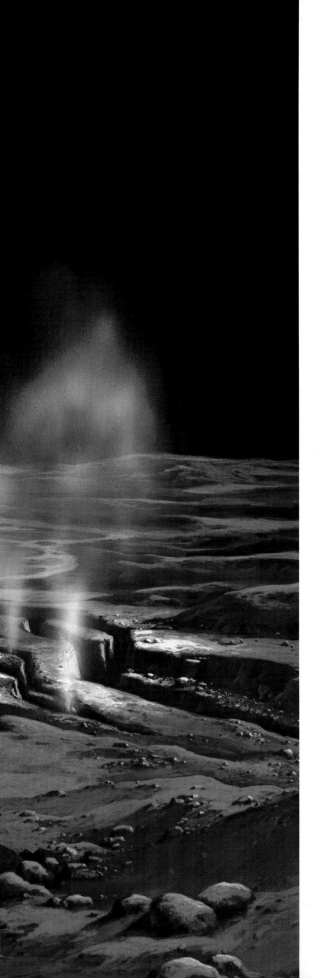

Fed perhaps by an underground ocean of sulfur, a distant volcano erupts into Io's airless sky. In the low gravity here, such fountains can shoot 300 km high and cover a distance as far as from New York to Detroit. Sulfur dioxide gas spurts from vents near a field of cold, blackened lava.

Io circles Jupiter in a wobbly orbit. While Jupiter tugs on Io from one direction, Europa and Ganymede sometimes tug on it from the opposite direction. This causes Io to weave first inward toward Jupiter and then outward. Its changing distances from Jupiter raise 100-meter ground tides. This means that Io's surface heaves up and down as if it were an ocean tide – to about the height of a redwood tree. The pumping action first stretches and then compresses Io's inner materials and so heats them. Is this heat the energy source for Io's volcanic activity? Or is the powerful electrical current from Jupiter the source? Or could it be a combination of the two? Whatever the source, Io's volcanoes seem to explode their sulfurous material with terrible force, propelling the particles to speeds as great as 3,600 kilometers an hour. This is several times the force of Earth's volcanoes, including Mount St. Helens and Mount Etna.

Io's sulfur snow

Astronomer Bradford A. Smith has said that Io looks something like a pizza. Voyager's images make its surface appear bright orange-red with white patches and smaller dark spots. But the color pictures are exaggerated to bring out detail. Strolling across Io we would be more likely to see colors that ranged from yellowish white to pale yellow-green. We would probably find foul-smelling sulfur dioxide snow. All around us volcanoes would hurl blobs and frozen flakes high into the sky. Io's ancient impact craters long ago were erased or covered (just as Earth's once-cratered surface

has been eroded over the ages). Io itself is as old as the other Galilean satellites. But, because of the volcanic snowfalls, some parts of Io's surface are only a few weeks old and it now seems doubtful that any part of its surface is over a million years old, so rapid is the change there.

Now and again we would come upon the remains of collapsed volcanoes and eroded slopes, the ghosts of an earlier crust. Voyager saw black spots, probably crusted-over lakes of lava that well up from below Io's top layer of frozen sulfur. These spots measure as high as 600°C, hot enough for sulfur to be molten, and much hotter than the surrounding surface of –145°C. From time to time we also would come upon vast lakes of molten sulfur. If we dug only a few kilometers down through Io's crust we might discover an ocean of liquid sulfur.

The Voyager spacecraft added an immense store of new data to our knowledge of Jupiter, but there is still much to learn. NASA is planning a new mission that will include dropping a probe into Jupiter's atmosphere (see illustration on page 264). The mission has been named, appropriately, Project Galileo.

Jupiter's enormous gravitational field has been speeding up our imaginary spacecraft ever since we came near this king of the planets. Now as we loop around we are flung off toward Saturn. Jupiter's gravity whip more than doubles our speed to about 135,000 kilometers an hour – about 50 times faster than the fastest jet airliners. Perhaps Saturn and its many moons will offer us a more promising environment for life. Or will they?

171

Saturn

A crescent Saturn looms large over the desolate, icy low hills of Rhea, one of the planet's seventeen or more moons. Rhea, about half the size of our Moon and more than half a million kilometers from Saturn, gleams bluish in the cold light of the distant Sun.

Saturn's magnificent ring system appears as only a thin line from Rhea because Rhea orbits Saturn at about the same plane as the planet's rings. But sunlight casts a wide ring shadow on Saturn – the dark patch that links the terminator and limb just above the thin ring line. When Galileo studied Saturn with his telescope in 1610, he saw the rings faintly as two knob-like shapes. He called them "ears."

Saturn's moon Tethys, about two-thirds the size of Rhea, is seen in crescent phase left of the terminator. The moon Dione is right of the ring. Although only slightly larger than Tethys, Dione appears considerably larger because it is closer to Rhea.

Rapid rotation causes Saturn to bulge a bit at the equator and flatten at the poles. The second largest planet in our Solar System, Saturn is a gas giant, like its larger neighbor, Jupiter. But because Saturn's clouds of ammonia snow are not so stormy, it's sometimes called the "quiet Jupiter."

est time in ancient Italy belonged
god of reaping, whom the Romans
Saturn. A symbol curved like his
— ♄ — represents the planet.

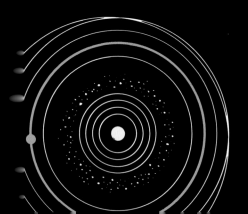

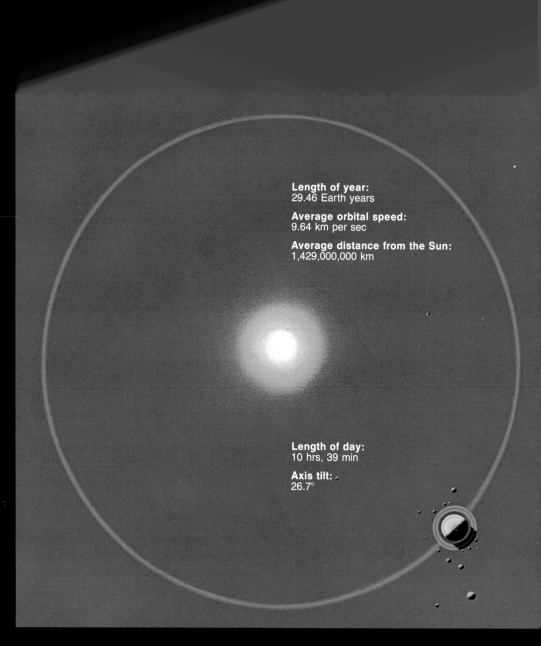

Length of year:
29.46 Earth years

Average orbital speed:
9.64 km per sec

Average distance from the Sun:
1,429,000,000 km

Length of day:
10 hrs, 39 min

Axis tilt:
26.7°

How far is it to the orbit of Saturn?
Imagine going all the way from Mercury
to Mars, through the asteroid belt, and on

Earth, since the poles never point our
way. Scientists know at least 18 moons
orbit Saturn. Titan, the biggest, holds

cy particles form the wide, thin rings that wheel above Saturn's clouds. Frigid at the top, the atmosphere grows thicker and hotter without ever reaching solid ground. Saturn radiates excess energy, maybe because heavier helium separates from hydrogen in the interior, turns to

drops, and sinks, the friction releases heat. Saturn's mass equals 95 Earths, but it takes up about 760 times as much room. That makes it the least dense gas giant— 70 percent as dense as water. Because of Saturn's fast spin and flat shape, gravity varies from poles to the equator.

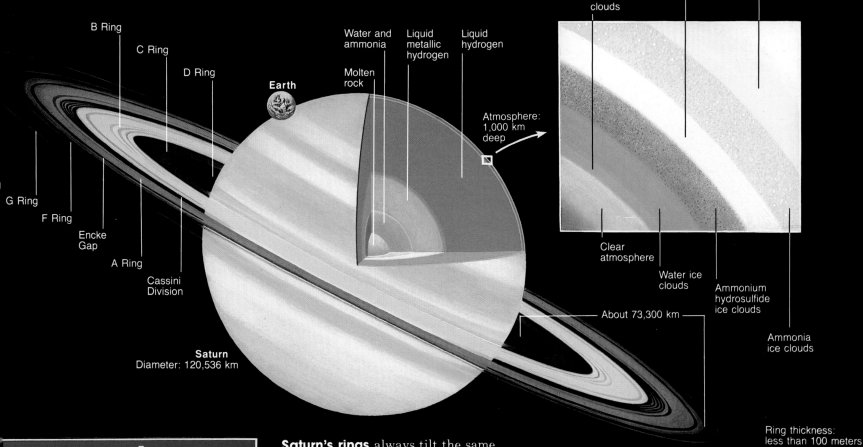

B Ring

C Ring

D Ring

G Ring

F Ring

Encke Gap

A Ring

Cassini Division

Saturn
Diameter: 120,536 km

Earth

Water and ammonia

Molten rock

Liquid metallic hydrogen

Liquid hydrogen

Atmosphere: 1,000 km deep

Cloud tops

Clear atmosphere

Water-ammonia clouds

Clear atmosphere

Water ice clouds

Ammonium hydrosulfide ice clouds

Ammonia ice clouds

About 73,300 km

Ring thickness: less than 100 meters

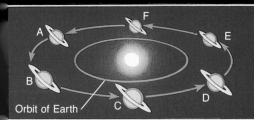

F

A

E

B

C

D

Orbit of Earth

Saturn's rings always tilt the same way, but from Earth the planet seems to tip its brim in different directions, depending on where it is. The orbit at left shows Saturn at points about five Earth years apart. At each position (below), we see the rings from a different angle.

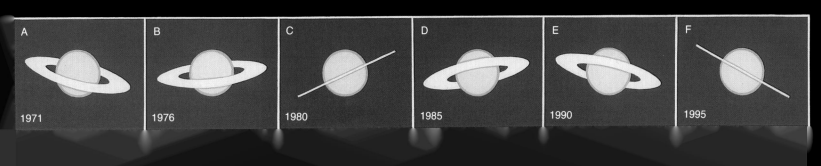

A 1971

B 1976

C 1980

D 1985

E 1990

F 1995

Saturn's parallel jet streams and brilliant halo of rings shimmer against the black sky in this enhanced Voyager portrait. Saturn's rings are gigantic. If you could take a space walk around the outer edge of the A Ring, going 25 km a day, you would be back at your starting place in 95 years.

A broken moon? Or a moon that never was? Astronomers do not yet know how Saturn's rings came to be. Full of icy particles (right), a ring seems to blur into a band of unearthly hail around the planet. One ring is so thick with rubble that almost no light can pass through it.

Our journey across space from Jupiter to Saturn has taken us a year and a half, traveling at an average 50,000 kilometers an hour. Saturn now looms ahead, a huge, misty, yellowish globe.

Of the planets, Saturn is second in size only to Jupiter. About nine Earths could line up along Saturn's diameter. This giant has about 95 times more mass than Earth, but it is loosely packed, more liquid than solid. It is the least dense of the planets, except possibly Pluto. Saturn's density is only 70 percent that of water — so low that if placed in water it would actually float.

Saturn's rapid rotation — 10 hours and 39 minutes — generates enormous centrifugal force and gives it a big bulge at the equator. There the diameter is 11,800 kilometers greater than from pole to pole. Saturn is the most oblate, or flattened, planet. Jupiter spins faster, but being denser is less "flexible," so it does not bulge as much.

Far from home

We are now so far from home that radio signals take about 90 minutes to reach Earth, and replies take just as long to get back. If we ask Earth's advice, we will get it in three hours. When we look back at the Sun, now 1.4 billion kilometers away, we see our

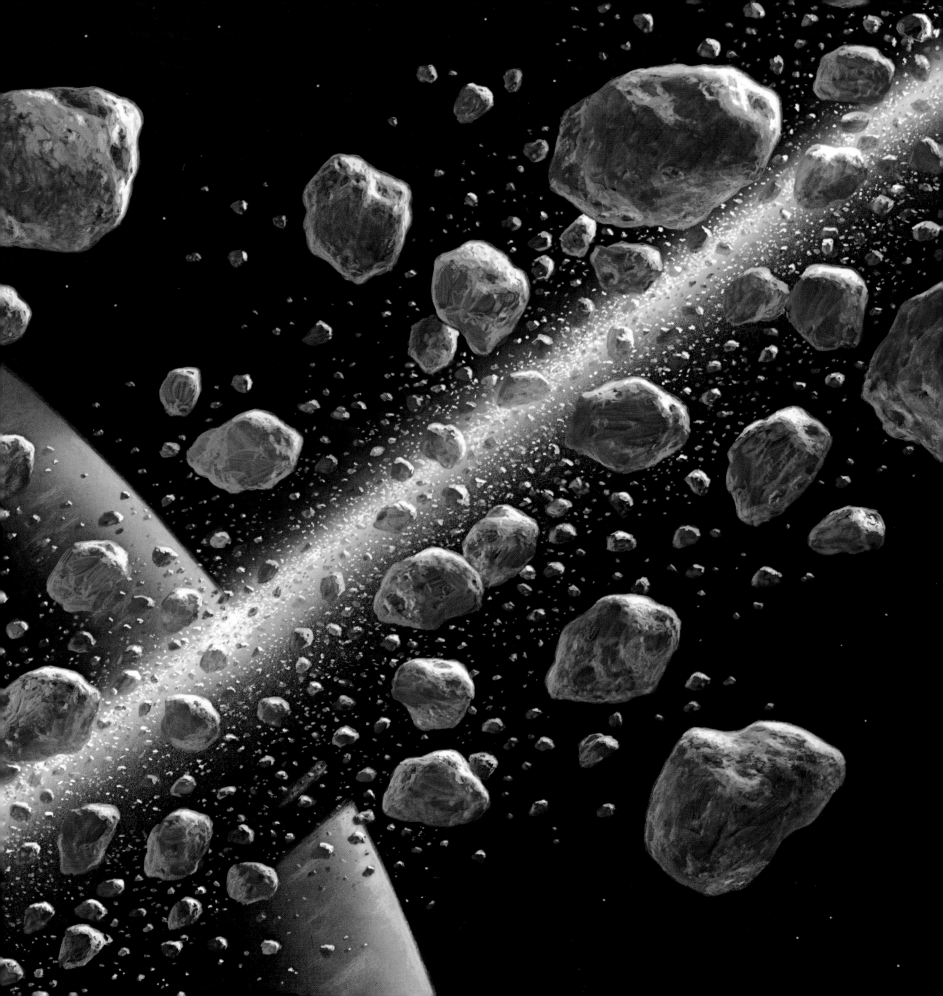

Thousands of individual multicolored ringlets, arranged like the grooves of a record, make up each of Saturn's main rings. NASA altered this Voyager picture of rings A, B, and C to emphasize the color contrasts. The inner C Ring and the Cassini Division appear dark blue. The dense and massive B Ring, brighter than all of the other rings combined, ranges in hue from golden orange to turquoise. More uniform, neutral tones mark the outer A Ring. These variations in color may reflect differences in the composition or size of the ring particles.

local star as a small disk of light thousands of times brighter than our full Moon. Saturn gets only a hundredth of the heat and light that we do on Earth.

At about a million kilometers from Saturn we meet the planet's powerful magnetic field. Much stronger than Earth's, this magnetic "bubble" pulses in and out with the changing force of the incoming solar wind. As with Earth and Jupiter, Saturn has radiation belts—charged solar wind particles trapped by the planet's magnetic field. If we entered these belts unprotected, the radiation would sizzle us.

The view up close

As we approach Saturn, the magnificent ring system fills the sky. From one outer edge to the other the rings stretch 960,000 kilometers, about two-and-a-half times the distance from Earth to the Moon.

We circle Saturn. Streaks of lightning tear through its yellowish-white cloud deck near the equator, where winds whoosh at nearly supersonic speeds. Floating crystals of ammonia ice give the upper clouds a yellowish color, but most of the atmosphere is hydrogen and helium with some methane and other gases.

The top of Saturn's atmosphere is much colder than the top of Jupiter's. We expected this, considering Saturn's much greater distance from the Sun. But farther down the temperature rises, more than we would expect from solar heating alone. Why this warming? Voyager unexpectedly found a "shortage" of helium in the upper clouds. It seems that much of the upper atmosphere's helium is sinking, warming the lower at-

mosphere by friction as it falls. Thus Saturn radiates nearly twice as much heat as it receives from the Sun.

Like Jupiter, Saturn is without a surface. If we could journey to the planet's center, the first half of the trip would be through a dense atmosphere mostly of hydrogen. Deep in this layer the pressure turns the hydrogen to liquid. Deeper still the layer of liquid hydrogen becomes so dense that it is best described as metallic. It is extremely hot here, about 10,000 kelvins. The center of the planet is probably an even hotter ball, about the size of Earth and made of water, ammonia, methane, silicon, iron, and other heavy elements, all in dense liquid form.

"Saturn has ears"

Galileo saw Saturn's rings through his telescope in 1610. He was probably the first person to do so, but he could not see them clearly and reported them as odd bulges. "Saturn has ears," he wrote. In 1656 the Dutch astronomer Christian Huygens identified the "ears" as a ring.

In 1675 the Italian astronomer Giovanni Cassini spotted a gap in the ring, now called the Cassini Division. The outer half is called the A Ring and the inner, brighter half, the B Ring. In 1850 other astronomers discovered the faint C Ring. But Saturn's ring story seems endless. In 1966 and 1969 observatory photographs showed an E Ring far out from the planet and a close-in D Ring. In September 1979, Pioneer 11 spotted the F and G Rings. The rings were named in order of discovery, not position. They start about 3,000 kilometers above

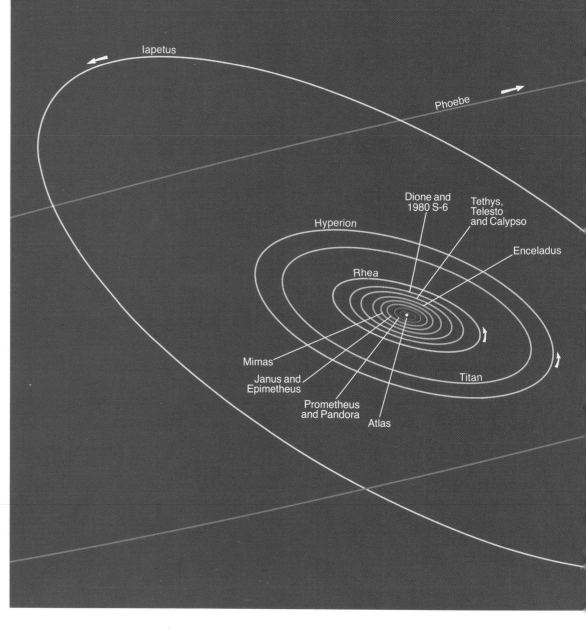

Saturn's moons form a mini solar system. Tiny Phoebe's tilted, retrograde orbit suggests it's a captured asteroid. A deep atmosphere makes Titan, the largest moon, a prime target for exploration. Although we know the orbits of only 17 moons, Voyager found evidence of others.

Saturn's equator and form a band some 420,000 kilometers wide.

We prepare to explore this sprawling system that makes the planet one of the most splendid sights in our Solar System. We start at D Ring, the innermost, and find it a wispy belt of thin ringlets. Saturn's rings are made up of thousands of such ringlets arranged much like the grooves of a phonograph record. The ringlets are composed of particles, mostly of water ice but perhaps some of rock, probably ranging in size from a dust speck to a house.

Next is the C Ring; from Earth it can barely be seen through a telescope. Between its sharp inner and outer edges the C Ring has two gaps and many light and dark bands. Within the outer gap is a bright ring.

Next is the broad B Ring, about 25,000 kilometers wide. It shines brighter than all the other rings combined and contains most of the planet's ring matter. It is so dense it casts a shadow on Saturn. Voyager also showed ghostly spokes, possibly clouds of tiny ice particles above the ring.

Rings with moonlets

The Cassini Division between the B and A rings was once thought to be empty. Voyager showed that it contains several bright ringlets separated by dark gaps in which there could be moonlets.

The outer region of the A Ring has several gaps, the largest one called the Encke Gap after the 19th-century German astronomer Johann Franz Encke. Within this gap are two bright and twisted ringlets, one nicknamed the Encke Doodle, and perhaps a moonlet.

The narrow F Ring lies beyond the A Ring like a pencil line only a few hundred kilometers wide. It encloses the main ring system in many strands of icy particles. Voyager discovered kinks in the F Ring. Two moonlets, Prometheus at the inner edge and Pandora at the outer, seem to keep matter in the F Ring from trailing off. Thus they are called "shepherd moons." A gravitational tug-of-war between them may account for the curious kinks.

Beyond F are the G and E Rings. They too are made mostly of small particles and are nearly transparent with diffuse boundaries. Widest and thickest of all the rings, the E Ring measures some 300,000 kilometers from inner to outer edge and several thousand kilometers from top to bottom.

How do planets get rings?

For centuries astronomers thought that Saturn was the only ringed planet. But Voyager found rings around Jupiter, and astronomers using a telescope spotted evidence of rings around Uranus. Could Neptune also have rings? Probably. How does a planet get rings? No one knows for sure.

Beyond a certain distance from a planet—a distance called the Roche limit—

Three of Saturn's large satellites—Tethys, Dione, and Rhea (from inside out)—hover just below Saturn's rings in this true-color image assembled from various Voyager pictures. Mimas, the planet's innermost major moon, casts a shadow on the butterscotch expanse of Saturn's clouds.

of matter can clump together and moons. But inside the Roche limit a net's gravity is too strong to let moons rm; particles there may remain as a thin veil of ring matter. So the question is: Where did the particles come from?

Just as the Sun had a great wheeling disk of icy and rocky material around it when it formed, Saturn and the other planets also probably had a disk of leftover matter, largely dusty ice, as the Solar System took shape 4.6 billion years ago. But the motion in Saturn's rings suggests they are younger than that; they may be debris from collisions between moons and comets.

Another possibility is that a moon or asteroid wandered in close to the planet and was broken into many pieces by the planet's gravity. Or perhaps early in the planet's development, two or more moons were forming, but smashed into each other and shattered, providing particles for rings.

We tour the moons
As our spacecraft nears the outer edge of the rings, we approach Saturn's major moons, one by one. We gaze about us at a miniature version of the Solar System. In this model, we can think of Saturn itself as the Sun, the ring particles as matter from

the original disk of ices, rock, and metal that wheeled about the Sun soon after it formed, and Saturn's moons as the planets.

Saturn has at least 18 moons. The largest is 5,150-kilometer-wide Titan. Six other major moons, made chiefly of water ice, range from 400 to 1,500 kilometers wide. Moonlets as small as 20 kilometers across sprinkle the system. Most of Saturn's satellites probably have gone through two major periods of bombardment—the first soon after they formed and were peppered by material forming the rings and moons, the second later on by moons colliding and showering nearby moons with debris.

Orbital dancing

Two of the small moons do a strange orbital dance unique in the Solar System. Called the "co-orbital satellites," they lie between the F and G Rings and circle Saturn about every 17 hours. One is 190 kilometers across, the other 120. Once every four years the inner moon overtakes the outer one. But the two orbits are only 50 kilometers apart, so there is no room to pass. Instead, just before collision, their gravitational attraction causes the two moons to switch orbits. The inner one, now in the outer orbit, slows down and lags behind as the other moves into the fast lane.

We reach the first major moon, Mimas, before we have left the outermost E Ring. Among its many craters is a giant crater 130 kilometers across and 10 kilometers deep — a third the size of Mimas itself. The crater must have been formed by a collision with a large asteroid. From its floor a central peak rises 6 kilometers high.

Pumping Enceladus

Next we pass Enceladus, one of the strangest little moons in the Solar System. Its surface has many kinds of terrain — old craters rounded over time, newer ones with sharp contours, broad plains, possibly even ice volcanoes. It appears to have been resurfaced many times by material welling up from inside. Whenever it lines up between Saturn and the moon Dione, gravitation squeezes and stretches Enceladus. This action may keep it geologically active and account for its varied surface. The activity may also create water volcanoes that spray out icy matter, giving the E Ring its

ice particles. The E Ring is somewhat brighter and denser near Enceladus.

Tethys, about 295,000 kilometers from Saturn, is a heavily cratered iceball about a third as large as our Moon. Its largest crater is 400 kilometers wide. Voyager photographed a deep, branching canyon about 1,000 kilometers long by 100 wide. Perhaps long ago Tethys froze from surface to center, and as the watery core froze, it expanded and cracked the brittle crust.

Dione is about the size of Tethys and nearly half again Tethys's distance from Saturn. It seems to be an ice-covered globe about 40 percent rock. Dione has surface markings that suggest a more active geological history than its twin, Tethys.

A midget moon

Dione has something else: a midget moon sharing its orbit. Why doesn't the big moon's gravity throw off the small moon? In the 1700s, French mathematician Joseph Louis Lagrange determined that a large moon's orbit could also hold one or more small moons, but only if they were about 60° ahead or 60° behind the large moon. In 1980 Voyager found Dione's little sidekick, and two others were later found sharing the orbit of Tethys — the Lagrangian moons, each 60° from the big moon.

Rhea is the second largest moon of Saturn. Four times the size of Mimas, it has a bright forward-facing hemisphere and a darker rear hemisphere with bright streaks. The streaks, as on Dione, may be signs of internal activity soon after the satellites were formed. Perhaps subsequent meteoric impacts erased the streaks on the

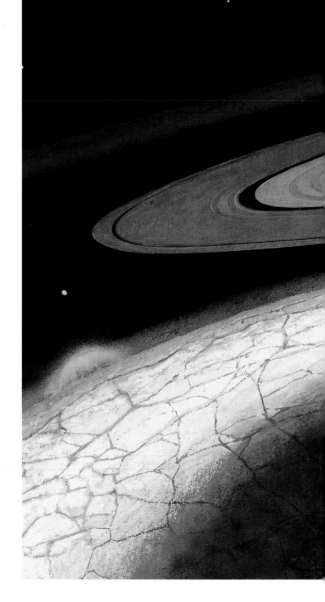

forward-facing hemisphere, leaving those on the more protected rear side.

Titan—king of Saturn's moons

Titan is special. Larger than Mercury, it has an atmosphere denser than that of any other moon in the Solar System. Telescope and spacecraft photographs show Titan glowing reddish-orange, which suggests an atmosphere mostly of hydrocarbons. But Voyager surprised everyone by finding that Titan's air is almost all nitrogen, like Earth's air. Some 200 kilometers deep, Titan's air includes argon, methane, and other gases present in Earth's early

atmosphere. Some scientists think Titan's atmosphere is evolving much as Earth's primitive air did some four billion years ago. "On Titan, we may have a snapshot of the atmospheric evolution that took place on Earth eons ago, with one major difference," said Rudolf Hanel, of the Goddard Space Flight Center. "It's so cold on Titan that nobody can imagine any life having formed there." Titan's cloud-top temperature is about −200°C. The atmosphere up there is composed of methane mixed with nitrogen. Deeper down the atmosphere is mostly nitrogen, much of it in the form of gas, but some of it as liquid droplets.

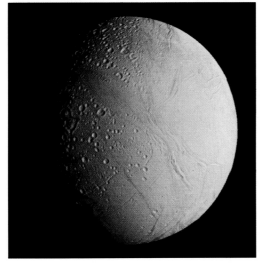

Brightest and smoothest of Saturn's icy moons, Enceladus (left and above) reflects nearly all of its light from the sun. Pure ice, more brilliant than snow, appears to form the tiny moon's highly reflective outer crust. Enceladus has many fresh, uncratered plains, suggesting that geologic activity continually reshapes its surface by smoothing out old cratered terrain. In one artist's view above, water from the moon's heated interior erupts through the icy crust and spills over the surface, forming plains. Such explosions might also blow ice crystals into space, replenishing Saturn's faint outer E ring (visible here as a blue band).

Titan's density is about twice that of water ice. So it probably is a bit less than half ice and the rest rock and metal materials.

Titan's surface must be a dark and gloomy place—an ice-covered world. Ethane and methane lakes may cover the equatorial regions. An eerie reddish sky hangs overhead, with floating clouds of methane ice crystals. And out of that sky a slow dusting of organic materials sifts down, covering part of Titan with a tarlike coating. Even so, Titan may be more like Earth than any other Solar System body. One day space probes may land on this remarkable moon to learn about the chemistry that gave rise to life on Earth.

Outer moons

Our journey continues past the minor moon Hyperion, a pear-shaped chunk apparently covered with dusty ice. We reach Iapetus, another mystery object of almost pure water ice. One side of this moon is white as snow and shines ten times brighter than the other side. The dark side may be covered with dust that falls on the moon from space, or perhaps with a thick, dark, inky substance that oozes from the interior of Iapetus.

Farthest of all the moons is lonely Phoebe, 13 million kilometers away from Saturn. Phoebe may be a captured asteroid, for it alone revolves backwards around the planet. As our spacecraft hurtles past, we recall one astronomer's poem:

> *Phoebe, Phoebe, whirling high*
> *In our neatly plotted sky,*
> *Phoebe, listen to my lay,*
> *Won't you whirl the other way?*

185

A World on Its Side
Uranus

As if twisted by a great cosmic hand, Uranus spins on a tipped-over axis. For 42 Earth years, one pole is bathed in sunlight, while darkness envelops the other. Once the planet has moved halfway around its orbit, the other pole will be aimed sunward for just as long. Voyager 2 scientists determined that, for some unknown reason, the dark pole is slightly warmer than the sunlit pole.

Uranus is so far from the Sun—19 times Earth's distance—that our local star appears only as a tiny disk of intense light. Uranus itself is bluish-green, probably because of methane in its atmosphere.

Voyager photographed 11 thin, coal-dark rings around Uranus, but it also observed flickering starlight, which indicates many other rings, or parts of rings. Though the particles in the rings are generally much larger than Saturn's, they are much sparser. Uranus's whole ring system contains less material than Saturn's Cassini Division, the spacecraft found.

The Voyager flyby also added 10 new moons to the 5 discovered from Earth. Here we view Uranus from a few hundred kilometers above the craters of outermost Oberon. With many different kinds of terrain, Miranda is the strangest moon,

Facts about
Uranus

Astronomers named the seventh planet for the old Roman god Uranus – father of Saturn and grandfather of Jupiter. Symbol: ♅, from a sign for the metal platinum.

Length of year:
84 Earth years

Average orbital speed:
6.8 km per sec

Average distance from the Sun:
2,875,000,000 km

Length of day:
17.3 hrs
Retrograde rotation

Axis tilt:
82.1°

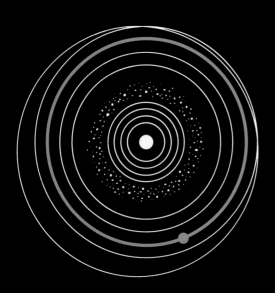

Like a lazy green giant with 15 gray dwarfs, Uranus and its moons tour our Solar System on their sides. As the planet orbits the Sun, its tipped-over axis stays aimed in the same direction – just as your foot keeps pointing forward on a revolving bicycle pedal. Here you see the Sun above the south pole; Uranus spins with one side in endless day, the other in night. But when Uranus reaches the top of our diagram, sunlight strikes it sideways, giving day and night from pole to pole. And a quarter-orbit later only the northern hemisphere sees light.

of the craft itself. They found that the rings are made up mostly of big boulders more than a meter wide. But instead of the icy coating that makes Saturn's rings shine so brightly, these rocks are black as coal. No wonder astronomers had trouble seeing Uranus's rings.

What is Uranus like?

As we close in on Uranus, we look down on a blue-green globe about 15 times the mass of Earth but only a fifth as dense. Methane in the upper atmosphere gives Uranus its color. But the murky air is mainly hydrogen with a little helium, a cold blanket of gases some 8,000 kilometers thick. Upper winds whoosh at 700 kilometers an hour, more than four times the high-altitude wind speeds on Earth. Unlike Earth's wind patterns, the winds of Uranus appear to blow in just one direction, the same way the planet turns.

Although each pole of Uranus stays in the dark for nearly half a century, the atmosphere cools so slowly that the dark pole is as warm as the one in the sunlight. The planet's temperatures are remarkably even: Uranus seems to transfer heat effectively from its poles to its equator. Like Jupiter and Saturn, Uranus gets hotter with depth. If we plunged through its frigid atmosphere we would probably splash into a deep sea of scalding water and ammonia over an Earth-size core of rock and metal that is hotter than the surface of the Sun— perhaps 7,000°C.

Instead we head outward to tour the rings and moons. The rings are much narrower than those of Saturn, and they con-

tain much less dust. If we could gather up all the boulders in all of Uranus's rings, we would have less material than we could sweep up in just the tiny F Ring of Saturn.

The innermost of Uranus's rings begins about 11,000 kilometers above the planet and spreads out over a width of about 2,500 kilometers. A gap nearly that width separates it from the next ring out. And from there we pass the main rings in rapid succession, since most of them are only a few kilometers wide. The outermost, known as the Epsilon Ring, is the widest at about 50 kilometers. By the time we reach its outer edge we are some 25,000 kilometers out from the planet, at the rim of a great black wheel of rings some 100,000 kilometers across, but only about a tenth the span of Saturn's rings.

Those marvelous moons

As with Saturn, the rings end about where the moons begin. Two shepherd moons perhaps 40 kilometers in diameter patrol the outermost ring. As they circle Uranus, they tug on the Epsilon Ring, fraying it into patchy strands and dense areas less than 30 meters thick. In all, at least ten small moons orbit Uranus, all of them discovered by Voyager.

But the five big moons we can see from Earth are the most fascinating—and the smallest and innermost of these, Miranda, steals the show. Though only 485 kilometers wide, Miranda is marked with craters, ropy patterns, deep valleys that turn at right angles, ridges that look like a racetrack, and cliffs that plunge 10 times deeper than the Grand Canyon.

Beyond Miranda we pass Ariel, about 1,160 kilometers across. On a surface grooved with wide, curving valleys and ragged canyons we spot several familiar-looking scars. They resemble the tracks of ancient glaciers back on Earth.

Next comes Umbriel, somewhat larger than Ariel. It looks dark, ancient, oddly quiet compared to the jumbled surfaces of the other four major moons. Then we notice a strange, bright circular feature. Scientists can't explain it, but its shape leads them to call it the "fluorescent Cheerio."

Now Titania fills our view ports, the largest of Uranus's moons at about 1,610 kilometers across. We see it as a splotchy sphere, split by long fissures that are edged in what looks like frost. Maybe it is; the cracks may have spewed out material that froze and fell alongside.

Outermost of the moons, Oberon orbits some 583,000 kilometers out from Uranus. It is nearly the size of Titania. A towering mountain thrusts up about five kilometers from a surface pocked with craters. Some craters look like they once filled up with something liquid that has long since hardened. Much of Oberon's terrain is splotched and streaked with light material.

All the major moons except dark Umbriel have shown us those whitish markings. And all except Umbriel showed terrain apparently pulled and squeezed by powerful forces sometime in the past. The best guess is that the planet's gravity was responsible for the geologic features we see on these moons. Why not Umbriel's too? We might land and try to unravel the riddle—but we have another planet to visit.

Last of the Giants

Neptune

Pale blue Neptune glows phantom-like against the Milky Way. The smallest and the last of the four gas giants, Neptune lies far out in "Solar Siberia," more than 30 times Earth's distance from the Sun. Here we see Neptune from its satellite Triton about 355,000 kilometers away. Directly behind us the Sun lights Triton's hilltops.

Of Neptune's eight known moons, Triton is the largest, slightly smaller than our own Moon. Its rocky surface seems to be covered with methane frost. Haze along the horizon is a sign of Triton's thin atmosphere of methane gas.

Along Neptune's equator a pronounced dark band separates the bright areas of the northern and southern hemispheres. These bright regions may be high-altitude clouds of methane ice crystals.

Because of its great distance from us, Neptune is hard to study, and we still have questions about the planet, but we have answered several. Six new moons have been discovered. In 1989, the Voyager spacecraft saw windstorms on Neptune like those on warmer Jupiter. As with Jupiter and Saturn, internal heat drives Neptune's turbulence. Scientists wondered if Neptune had rings. It does—narrow, faint, and dark.

Facts about
Neptune

An ocean god for an ocean-colored
planet—Neptune, Roman god of the sea.
Astronomers use his fishing spear, the
trident, for the planetary symbol: ♆

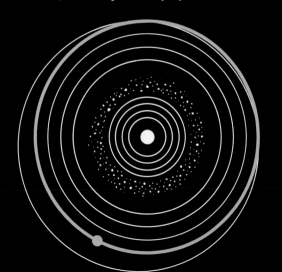

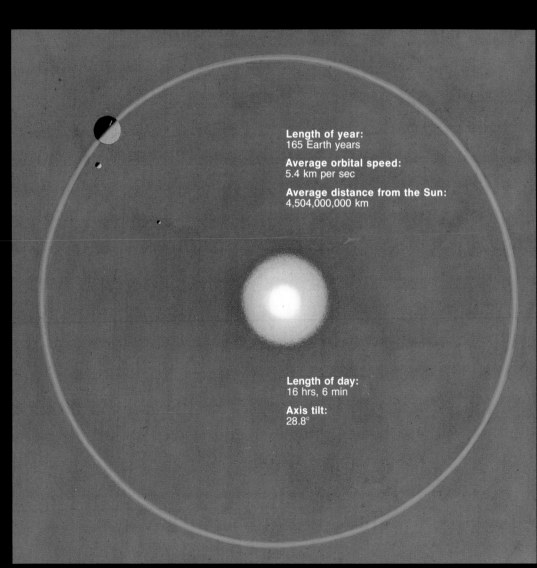

Length of year:
165 Earth years

Average orbital speed:
5.4 km per sec

Average distance from the Sun:
4,504,000,000 km

Length of day:
16 hrs, 6 min

Axis tilt:
28.8°

Neptune creeps in its path. Discovered
in 1846, the eighth planet will not
return to the position where scientists
first saw it until 2011. Neptune orbits
slowly because it is far from the Sun. We
cannot see it without a telescope. In size
and color Neptune seems to be a twin to

Uranus. But Neptune does not spin on
its side; its axis tilts only a little more
than Earth's. The orbits of two moons,
Triton and Nereid, crisscross the equator
diagonally. Between Triton, the largest
satellite, and the planet itself, six small,
recently discovered moons also orbit.

The odd couple—Neptune's puzzling moons (below) don't act like any other planet's. Nereid, only 340 km wide, follows the most elliptical path of any satellite. It needs an entire Earth year to complete one orbit. Triton, a giant about 2,700 km across, takes less than six days to circle Neptune—backwards. No other large satellite moves opposite to the direction in which its planet spins. Some scientists think ancient Neptunian satellites might have collided, throwing some moons out of orbit and leaving these two in their strange courses.

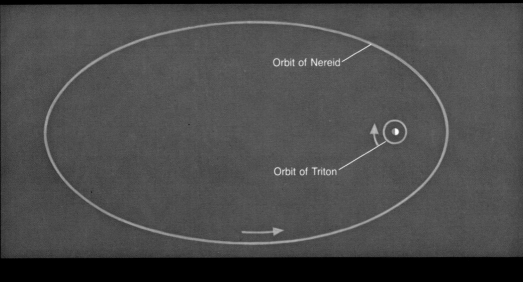

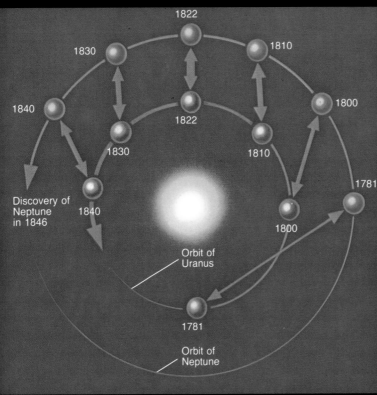

The newly discovered planet Uranus was behaving in a strange manner. Instead of orbiting the Sun smoothly, it sometimes sped up and at other times slowed down along its path. Why?

The Solar System grows

The year was 1841. John Couch Adams, a college student in England, read a report about Uranus's odd motion. Was it possible, as astronomers were suggesting, that an unknown distant planet was tugging at Uranus and so causing it to move unevenly? Adams was fascinated and decided to study the matter after he graduated from Cambridge University.

By September 1845 Adams had calculated the orbit of the suspected planet and pointed out the place in the sky where the planet should be found. He presented his work to the Royal Observatory in Greenwich, but no one took him seriously. Meanwhile other astronomers had joined the search. In France, Urbain Leverrier also worked out the new planet's orbit. On September 23, 1846, the planet was sighted in the place where Adams and Leverrier had said it would be. It was the first time a planet had been found by mathematics. Adams and Leverrier share the credit.

Before Neptune's discovery the known Solar System stretched out to almost 20 times Earth's distance from the Sun. The new planet increased the span to 30 times.

Rings around Neptune

Through a telescope on Earth, Neptune appears as a faint bluish-green disk two-thirds the size of Uranus. Actually it is

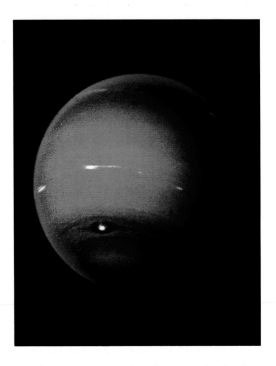

nearly the same size but much farther away. Four Earths could be lined up across Neptune's diameter. The planet is about 17 times more massive than Earth. A Neptune year is 165 Earth years. And, as Voyager discovered by measuring radio emissions, a Neptune day lasts 16.1 hours—shorter than previously thought.

Soon after Neptune was discovered, at least two astronomers reported sighting rings. In recent years when rings were found around Jupiter and Uranus, astronomers again began wondering whether Neptune might also have rings. But telescope observations from Earth proved puzzling. Then Voyager found three prominent rings. Embedded in the outermost ring are arc-like clumps of ring material. No one, as yet, can explain the clumps.

Astronomers now regard Neptune and Uranus as twin planets, somewhat alike in size, density, mass, and rotation. In most physical respects except size they closely resemble Jupiter and Saturn.

What is Neptune like inside?

Before Voyager we could only make guesses based on the planet's mass, density, and size and on photographs taken through telescopes. Now astronomers believe that Neptune's upper cloud layer is made of methane ice. A lower cloud layer is perhaps hydrogen sulfide ice. The cloud-top temperature is a frigid –200°C. Neptune's blue color comes from the scattering of sunlight—as with Earth's atmosphere.

Neptune's core seems to be a ball of molten rocky material slightly larger than Earth, with a temperature around 7,000°C.

We now know that Neptune emits more energy than it receives from the Sun. So do Jupiter and Saturn, each for different reasons. Where does Neptune's excess heat come from? Scientists think that its source is the cooling of the planet's core.

Neptune's largest moon, Triton, astonished Voyager scientists. One called it "a world unlike any we've ever seen." With a surface temperature of –235°C, Triton is the coldest body yet explored in the Solar System. Its crust is frozen hard as stone. A south polar ice cap is mostly frozen nitrogen. Triton displays activity known as ice volcanism—geyser-like plumes of nitrogen burst through weak spots in the crust, rising some eight kilometers above the surface. Perhaps solar energy causes the eruptions but no one knows for sure.

Pluto

The planetary sentinel stationed at the distant border of the Solar System, Pluto ranges between 30 and 50 times farther from the Sun than Earth is. We view Pluto from its icy and pitted companion Charon. Only recently discovered, Charon is believed to orbit about 19,000 kilometers from Pluto's icy crust. Here Pluto, in thin crescent phase, is dimly lit by the distant sun, the bright point of light at lower right, as the night side of the planet glows feebly in light reflected from Charon.

Pluto is not much more than twice the size of Charon, which makes Charon the largest moon in the Solar System in relation to its planet. Both are so small and so far away from us that they are hard to study with even the best telescopes. Both planet and satellite seem to wear a coating of methane ice, but astronomers do not know what the interiors are like.

Is it possible that Pluto or Charon are remains of a shattered moon once belonging to Neptune? Or might Pluto and Charon be lumps of matter almost unchanged from the time the Solar System formed?

Because Pluto and Charon are so close and similar in size, some astronomers consider them a double planet rather than a planet with a satellite.

Facts about
Pluto

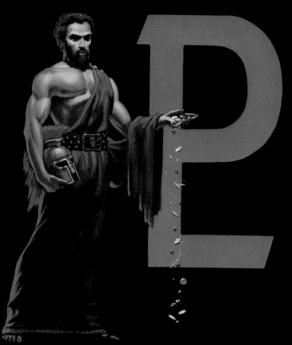

Greek god of wealth, Pluto ruled the dark underworld of myth. Now the darkly lit outermost world bears his name. The "Pl" of Pluto is the planet's symbol: ♇

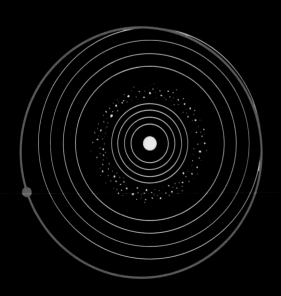

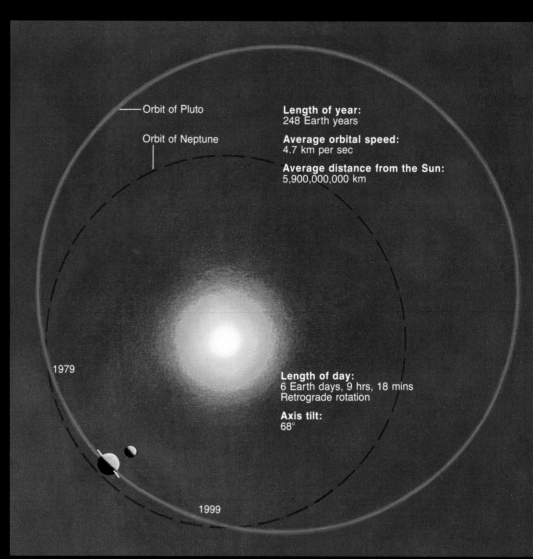

—— Orbit of Pluto

Orbit of Neptune

1979

1999

Length of year:
248 Earth years

Average orbital speed:
4.7 km per sec

Average distance from the Sun:
5,900,000,000 km

Length of day:
6 Earth days, 9 hrs, 18 mins
Retrograde rotation

Axis tilt:
68°

Last and least, tiny Pluto follows the most elliptical orbit of any planet. It ranges between 4.4 and 7.4 billion km from the Sun. From 1979 to 1999 Pluto actually lies closer than Neptune. The last time that happened, about one Plutonian year ago, George Washington was in his boyhood. Will Pluto ever hit Neptune? No; Pluto's orbit tilts, missing Neptune's. Until about 1991 twice-weekly eclipses between Pluto and its nearby moon Charon (right) will enable astronomers to learn their size and composition and to map one side of each.

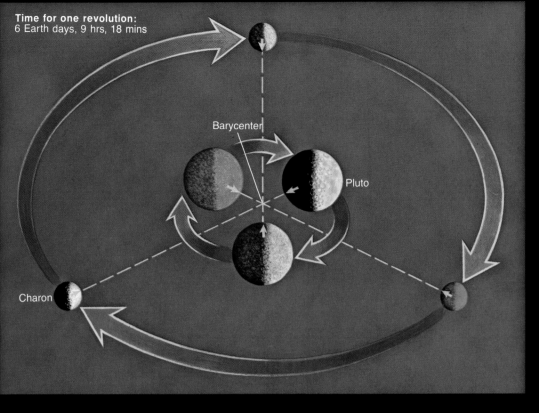

Time for one revolution:
6 Earth days, 9 hrs, 18 mins

Barycenter

Pluto

Charon

Pluto and Charon appear to spin like two twirling dancers, one light and one heavy. Just as the lighter dancer whirls in the wider circle, so Charon revolves farther from the balance point, or barycenter, than more massive Pluto. Each keeps the same side to the other.

A riddle—Gravity, maybe 4 percent of Earth's. Density, less than ice. Mass, about one-sixth of our Moon. Core, unknown. Pluto's small size (below), low mass, and odd orbit make some scientists ask if this seeming ball of frozen gases really is a planet. But if not, what is it?

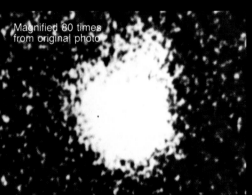

Magnified 80 times
from original photo

Charon

Pluto

Earth

Pluto
Diameter:
2,300 km

Discovery of Charon—In 1978 a photo like the one above puzzled astronomer James Christy. Why was Pluto bulging?

Then he decided he had found a moon, about 1,400 km wide, the Solar System's largest in relation to its planet.

Pluto to Charon: 19,000 km

Earth to Moon: 384,000 km

Two pictures of the same star field, taken days apart, led to Pluto's discovery in 1930. Clyde Tombaugh used a device that flashes quickly from one such photograph to the other. If anything has moved—such as a planet—it blinks. He flashed pictures for months before Pluto winked at him.

Stuck in eternal moonrise, Charon hangs in the Plutonian sky as if frozen by the −230°C cold. A pinpoint Sun, intensely bright, glints on a field of methane ice. Since Charon and Pluto spin locked face to face, the big moon seems always to hover over the same place.

Something still seemed wrong. Not only was Uranus not staying on schedule as it orbited the Sun, but Neptune also seemed to be out of step. Could a ninth planet be tugging on Uranus and Neptune from the far reaches of the Solar System? In the early 1900's, one astronomer who thought so was Percival Lowell. He devoted his life to looking for what he called Planet X.

Lowell did not live to see his planet. But in 1930, Clyde Tombaugh, a young assistant at the Lowell Observatory in Arizona, was continuing the search. After months of work he found Planet X. An 11-year-old English girl was the first to suggest that this dark and gloomy place be named Pluto, after the Roman god of the underworld.

A cold little world

Although it may have a thin atmosphere, the planet seems to be only a snowball of frozen gases, not massive enough to affect either Uranus or Neptune. Probably their orbits were measured inaccurately. It was thus a great coincidence that Pluto was where Lowell had predicted it to be.

From Pluto, the Sun is a tiny point of light but, for the next few decades, a point about 440 times brighter than the full Moon from Earth. Temperatures hover only a little above *absolute zero,* the point at which there is no heat at all and molecules stop moving entirely.

In 1978 James W. Christy, of the U.S. Naval Observatory, discovered that Pluto has a moon, which he called Charon. Until about 1990, astronomers can watch frequent eclipses between Pluto and its moon, a chance that comes along only every 124 years, when their orbits line up. Measurements of these eclipses should tell us the bodies' sizes and what they are made of.

Is Pluto a true planet?

Some astronomers ask the question. Pluto's orbit is unusually stretched-out; it tilts sharply from the orbits of the other planets, and it crosses Neptune's path. Nowhere else in the Solar System does this happen. Pluto's small size—about the same as our Moon—also is unusual. And finally, as a world of frozen gases, it seems to resemble some of the outer planets' moons.

Could Pluto be an escaped moon of Neptune? Some astronomers think so. Long ago a large planet may have swept in close to Neptune. Its gravitation could have pulled Pluto away, flinging it into its long elliptical orbit around the Sun and breaking off a piece that became Charon.

But what happened to the intruder planet? Is it possible that the mystery world might be Planet Number 10 in the Solar System? Some astronomers speculate that the near miss with Neptune threw it into an orbit far from the Sun, and that it is out there somewhere, too dim to see from Earth. But so far there is no evidence for it.

In 1977 at Mount Palomar Observatory, Charles Kowal discovered an object out beyond Saturn. At first he thought it might be Planet 10, but it turned out to be a small world like the outer moons of Saturn and Uranus. Called Chiron, this minor planet also has a stretched-out orbit, one that takes it across Saturn's path.

It is tempting to think that both Pluto and Chiron are escaped moons but more likely Pluto, Charon, and Chiron are leftovers from the Solar System's formation some 4.6 billion years ago. For some reason the giant planets never scooped them up. They could be museum relics, and in their frozen landscapes may lie secrets from the age when the Sun and planets took shape.

The Comets

A giant beam shining out of the dark, a long-tailed comet passes Earth and our Moon on its way around the Sun. Sunlight and the solar wind keep a comet's tail always pointed away from the Sun so, when the comet leaves the inner Solar System, it will speed away tail first.

Here both Earth and the Moon are lighted not by the comet—which gives off little light of its own—but by the Sun.

People of old were terrified when a comet came this near Earth. They believed that comets brought earthquakes, wars, and famine. In 1910 people feared that the gases of Halley's Comet would poison them.

A comet's solid nucleus is hidden within its surrounding cloud of gas and dust. The cloud is blown out into two tails—a yellow one made of tiny particles that have been freed from the ice-ball nucleus as the ice heats and vaporizes, and a bluish one made of electrified gases including carbon monoxide, water vapor, and nitrogen.

Some comets travel through the Solar System and do not return for a thousand years or more. Where do they come from? Where do they go? "Comet dust," analyzed by scientists, seems to be more than 4.5 billion years old: Comets are part of the original matter of our Solar System.

Most comets are named after their discoverers. Ikeya-Seki honors Japanese amateur astronomers Kaoru Ikeya and Tsutomu Seki. Halley's Comet recalls England's Edmund Halley. Caroline Herschel of England discovered eight comets between 1786 and 1797.

Tangling the Solar System with orbits, comets zip past the Sun, each at a frequency called its period. Encke has the smallest orbit, thus the shortest period: 3.3 years. Kohoutek, seen in 1974, may not be back for a million years or more. First it may loop far out toward

the Oort Cloud, a vast region of widely separated comets. Some comets we see originated there. If one breaks up in its travels, the pieces string out along its orbit, in a comet group. One group, the sungrazers, passes within a Sun's diameter of the solar surface.

A comet begins as a nucleus of frozen materials. When near the Sun, the comet heats enough that some of these materials turn into gases and form powerful jets that carry away solids as dust. The jets slightly change the motion of the nucleus. The gas and dust form distinct tails.

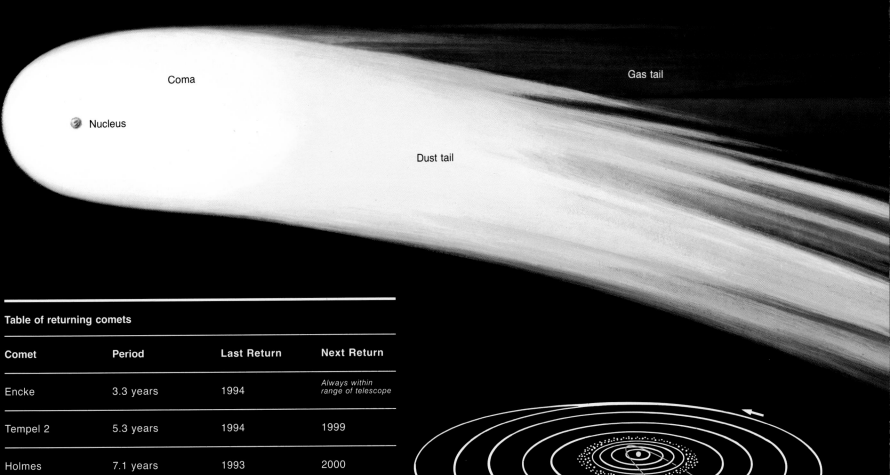

Coma

Nucleus

Gas tail

Dust tail

Table of returning comets

Comet	Period	Last Return	Next Return
Encke	3.3 years	1994	*Always within range of telescope*
Tempel 2	5.3 years	1994	1999
Holmes	7.1 years	1993	2000
Faye	7.4 years	1991	1999
Halley	75-76 years	1986	2061

Orbit of Halley's Comet

A comet's period is like a planet's year: the time it takes to complete one orbit. But each time a comet approaches a planet, its orbit may be changed by the planet's gravitational pull. Many comets are eventually pulled close to the orbit of Jupiter—sometimes they stay there.

Halley's Comet returns to our view about every 76 years on a long orbit that tilts some 18° from the Solar System's orbital plane. In 1948 the comet began its latest journey from beyond Neptune, passing the orbits of seven planets before rounding the Sun in February 1986.

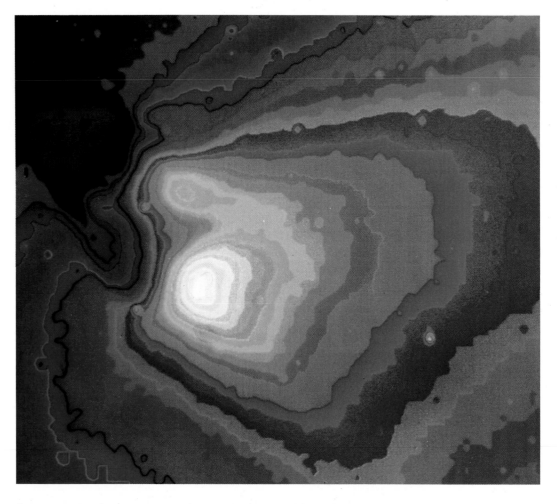

A 1986 computer-colored image of Halley's Comet, taken by the European space probe Giotto, shows the nucleus as a dark green blob at upper left. The white bull's-eye area is a dust jet thrown off from the nucleus. In 1965 comet Ikeya-Seki (opposite) lighted the sky for months.

Some people of ancient times believed that a comet blazing through the sky was the soul of a hero or a king on its way to heaven. Others thought that comets were messengers of widespread disease and doom. They called comets "terrible stars" or "death-bringing stars." Even in recent times, people have feared exposure to the poisonous gases of comet tails. They feared that if Earth passed through a comet tail all life would surely die. When Earth swept through the splendid tail of Halley's Comet in 1910 (with no noticeable effect), enterprising merchants sold "comet pills" as protection against disaster.

One present-day writer has described comet tails as "about as close to nothing as something can get." And another has said that they are "a visible nothing." They are like "dirty snowbanks," said astronomer Fred L. Whipple, collections of rocky matter and dusty ice.

What comets are made of

Actually, as recent spacecraft flybys of Halley's Comet discovered, they may be like fluffy rubble heaps. The solid part of a comet is called the *nucleus*. Halley's elongated nucleus measures about 15 by 8 kilometers and is coal black, perhaps because it contains carbon. It appears to be a heap of icy matter wrapped in a crust of dark matter, possibly rock dust. But, when near the Sun, its surface temperature is a surprisingly warm 330 kelvins. Powerful jets of gas and dust stream from openings in the surface.

When a comet sweeps in near the Sun, heat boils off the ice, making a cloud. This cloud encloses the comet in a dusty cocoon called the *coma*, which may reach out great distances.

When we look at a bright comet, we can sometimes see that its tail has two distinct parts. Each is pushed out from the coma by a different process. The yellow part—the *dust tail*—gets its color because sunlight reflects off the dust grains. It is smooth and curved. The pressure of sunlight photons pushes out this part of the tail. The streaming solar wind pushes out electrified atoms (ions) from the coma and forms the *gas tail*, also known as the *ion tail* or *plasma tail*. Some of these ions are carbon monoxide, and glow with a blue light. The ion tail may contain kinks or twists. Because of this pushing action of the Sun, the tail of a comet is always kept pointing away from the Sun.

As a comet nears the Sun, the tail stretches out behind for many millions of kilometers. As the comet loops around the Sun and then speeds away in the same direction from which it came, its tail is ahead of it rather than behind. Coming or going, the tail points away from the Sun and is so thin that we can see stars shine through it.

Parent molecules

When the frozen matter of the nucleus turns directly from a solid to a gas—a process called sublimation—the gas soon disintegrates into what scientists call "breakdown products of the parent molecules." The coma and ion tail of a comet are made up almost entirely of breakdown particles, since the parent molecules disinte-

As a comet dives inward, solar heat boils off bits of its nucleus; its tail grows, always pushed away from the Sun. On the outbound trip, the tail fades into space. A weakened comet may fall apart. When Earth passes through an orbiting cloud of debris, we see a bright meteor shower.

Space bombardment: In July 1994, 21 large fragments and many smaller ones of the comet Shoemaker-Levy 9 slammed into Jupiter and exploded. The Hubble Space Telescope photographed the dark impact scars in the upper atmosphere, huge areas of dark particles and gases.

grate deep inside the coma, close to the nucleus. Results from the recent comet fly-bys confirm that the most common parent molecule is water. But there may be some "dry ice," or frozen carbon dioxide, and other parent molecules in comets as well.

Where do comets come from?
Astronomers have studied enough comets to know that these visitors from afar are members of the Solar System, but they are very distant members.

Comets seem to have a "home ground" in the farthest reaches of the Solar System, almost 7 trillion kilometers away from the Sun—about 45,000 times farther from the Sun than Earth is. This home ground is a huge region called the Oort Cloud, after the Dutch astronomer Jan H. Oort. Oort's idea is that some 100 billion comets swarm there. Most astronomers think that this enormous collection of comets is left over from the time the Sun and planets were formed out of primordial matter some 4.6 billion years ago. The dust in comets may be much like the dust grains in the Horse-head Nebula, for instance. The frozen gas in the comet nucleus may be similar in chemical composition to the gases found in the dark clouds of the Milky Way.

The leading theory on the origin of comets holds that they once orbited near the edge of the Solar System. Eventually millions of them combined to form Uranus and Neptune, and billions more were flung out into the distant deep freeze of the Oort Cloud by the planets' gravitation. An alternative idea is that the comets did not form within our Solar System at all, but rather

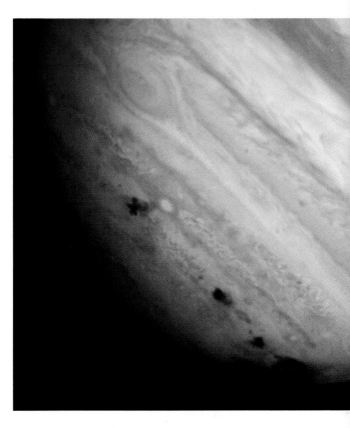

that they were captured by the Sun from far outside it in deep space.

In any case, the gravitational tug of a passing star can snatch a comet from the Oort Cloud and fling it into a cigar-shaped elliptical orbit that brings it close to the Sun. Some comets stay in this path, with one end of the long orbit near the Sun and the other end far out in the Oort Cloud. These comets can take millions of years to complete one orbit. For every comet snatched out of the Oort Cloud and sped toward the Sun by a passing star, there probably is another that is gravitationally whipped away and lost to deep space.

June 30, 1908—An explosion as powerful as a 40-megaton atomic bomb rips the morning sky over a Siberian wilderness— long before atomic bombs existed. It levels 2,000 sq km of forest in the Tunguska River valley, yet digs no crater. People hear it 1,400 km away; just before, reports say, a fireball streaked overhead. The Tunguska debate still goes on—was it a meteoroid? An asteroid? UFO cultists suggest an alien spaceship. Some scientists think they know: A small comet, made of loosely packed dust and ice, blew up from its collision with the atmosphere.

Comet families

As it sweeps in among the planets, a *long-period comet* may be disturbed by the gravitation of Jupiter or one of the other giant planets. Its orbit may then become much shortened, one end near Jupiter, say, and the other end near the Sun. These *short-period comets* are members of the inner Solar System. Encke is the comet with the shortest known period. It orbits the Sun in only three years and four months. Jupiter has captured several comets and thus has its own *comet family*. While Jupiter may tug a long-period comet into a smaller orbit, it is just as capable of flinging a short-period comet into a larger orbit—but always around the Sun, not the planet.

Halley's Comet

After studying the reported appearances of many comets, around 1700 Edmund Halley said that the comets sighted in 1531, 1607, and 1682 were not different comets but the same one that had returned again and again at intervals of about 76 years. The earliest recorded sighting was in 240 B.C. by Chinese observers who recorded the appearance of a "broom star." As a boy of 14, Julius Caesar saw the comet in 87 B.C. In Europe many people blamed it for causing the plague in the years 530 and 684. England's King Harold took the comet's return in 1066 as a dreaded sign of doom. Then, sure enough, the invading Normans defeated him in battle. The ruthless Mongol warrior Ghengis Khan saw the comet in 1222 and, the story goes, claimed it as a sign for him to conquer the world. So he led his army across Asia and slaughtered hundreds of thousands of people.

The famous comet, which has so far made 30 recorded visits to the Sun, was named in honor of Edmund Halley. After many calculations and long discussions with Isaac Newton about comet orbits, Halley had predicted the comet's return again in 1758. It appeared on schedule, but Halley had died 16 years before.

In our own century, Halley's Comet put on a grand show in 1910, when it passed only 22 million kilometers from Earth. During the most recent visit in 1985-86, the comet was much farther away at its closest approach to Earth—about 63 million kilometers—so people could not see it well. But this time an international committee of spacecraft went out to meet it.

The Soviet Union sent two Vega crafts, which found the dense cloud of dust that envelops the comet's core. The Japanese spacecraft Suisei, when much farther away, was able to report that Halley's rotates every 52½ hours. It observed that hydrogen, presumably from water vapor broken up by ultraviolet sunlight, forms much of the atmosphere around the core.

Giotto, a spacecraft sent by the European Space Agency, got closest. It shot within 605 kilometers of the solid nucleus on what some scientists called a "suicide mission." Comet dust struck the craft at speeds 70 times that of a bullet and badly battered it. But Giotto managed to send back enough images of the nucleus and other data to keep research astronomers busy for years. Since scientists believe that the chemical composition of Halley's Comet has changed little since the Solar System formed, they expect to discover some important clues about the Solar System's origins.

Split comets

When a comet loops around the Sun and heats up, jets of gas may spurt out of its nucleus and act like the thrusters of a rocket. This may set the nucleus spinning, or speed up the comet, or slow it down, and so change its orbital path. Then the comet may be off schedule. Sometimes the nucleus splits into two or more pieces. This may be caused by the force of the gas jets, or by heat and gravitational force when a comet nears the Sun. If the fragments stay in orbit together, they are called a *comet group*. We know of about 20 split comets. Sometimes a comet does not survive; it breaks up completely and vanishes.

The sungrazing comets

Some comets sweep in so close to the Sun that they are called *sungrazers*. We know of more than a dozen. Some come as close to the Sun as the Sun is wide. Others crash into it, as did Comet Howard-Koomen-Michels in 1979, one of five sungrazers discovered by the U. S. Navy telescope SOLWIND in orbit aboard a U. S. Air Force satellite. Sungrazers have long periods— from 500 to more than 1,000 years. Together they make up a comet group that may be the remains of an ancient, giant comet that broke up. They head in toward the Sun from below Earth's orbit and whip around it in a direction opposite from the planet's motion around the Sun.

One of the most famous of the sungrazers is Comet Ikeya-Seki, officially called 1965

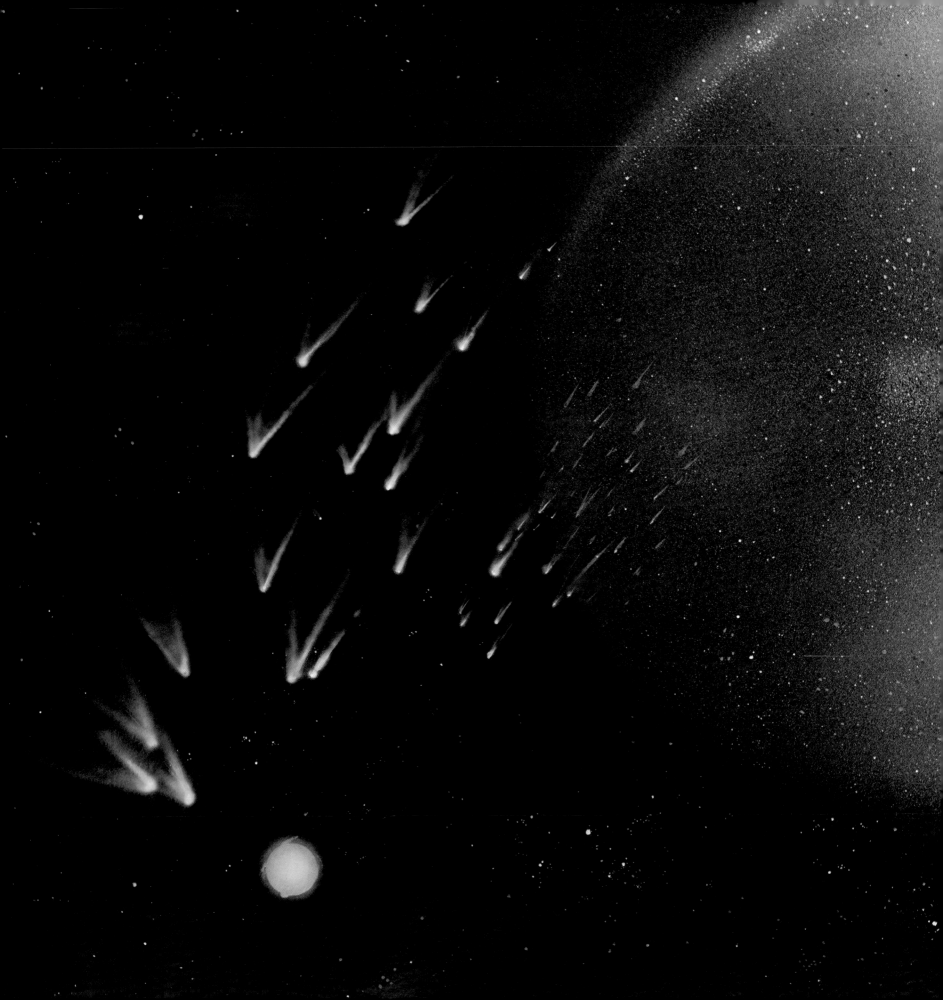

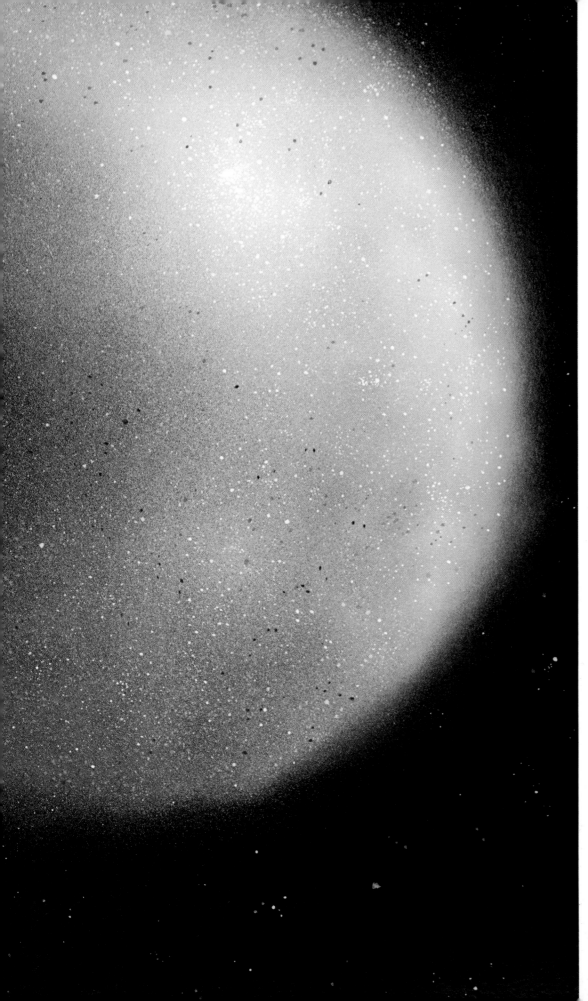

VIII. That year Ikeya-Seki put on a show which millions of people saw. When near the Sun it was visible in broad daylight.

Soon after Ikeya-Seki's closest approach to the Sun, the nucleus broke into pieces, split by the heat and gravitational force of the Sun. The same thing had happened to another brilliant sungrazing comet, 1882 II, which has been called "a near twin of Ikeya-Seki." The nucleus of the 1882 comet broke into four pieces; two survived.

What happens to comets?

Each time a comet rounds the Sun and flares out with its coma and tail, it loses part of its frozen material to space. Gases stream into the coma and tail.

Dust from comet tails drifts off into orbit around the Sun. Pieces the size of sand grains form long swarms sharing the same orbit. At regular times each year, Earth passes through some of these swarms. Then the fragments rain down through our atmosphere, where they heat up. They glow like spacecraft on reentry; they are the familiar meteor showers. The surfaces of some comet nuclei may lose their ices, leaving an outer crust of rock dust. Such nuclei would become objects resembling asteroids. Thus, some of the objects that astronomers call asteroids may actually be the dead nuclei of comets.

So far as we know now, the outer boundary of the Oort Cloud marks the limit of the Solar System. Beyond lies interstellar space. There distances are so great that we must shift the gears of our minds to imagine—if we can—the immense size of our local galaxy of stars.

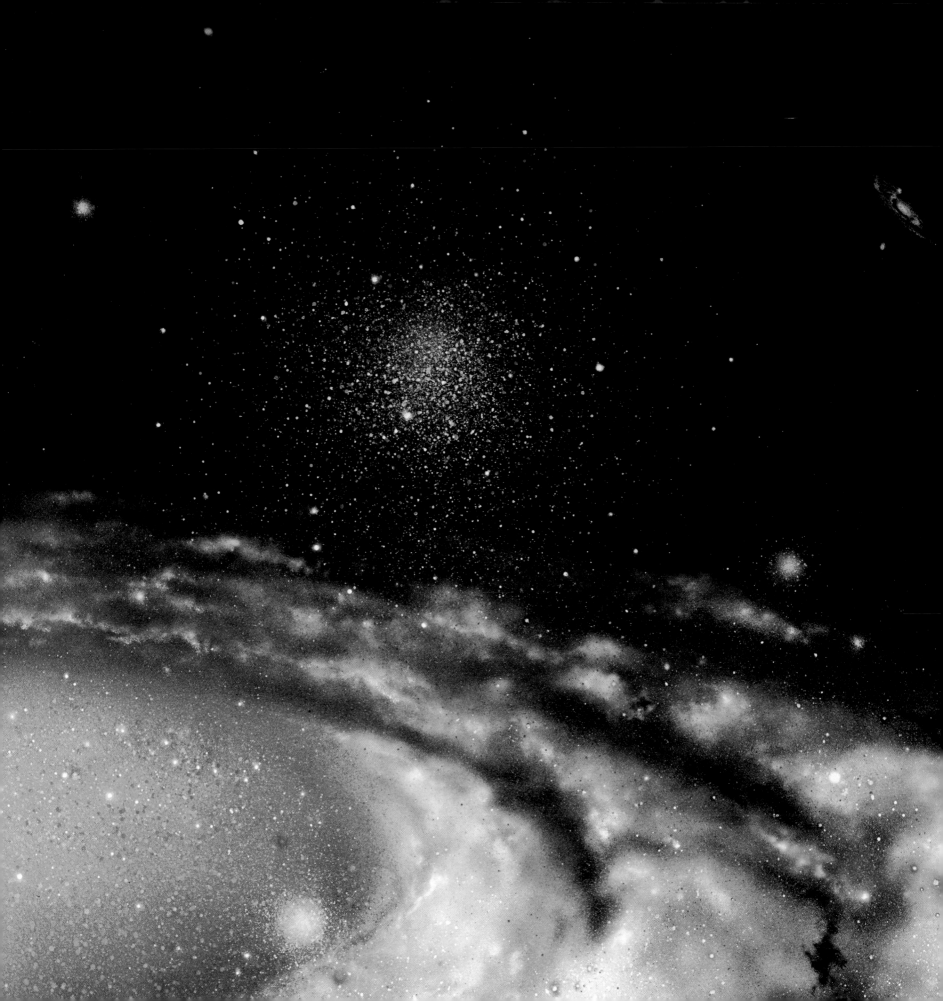

To the Stars & Galaxies
Deep Space

Our home galaxy, the Milky Way, spins its spiral arms of gas, dust, and billions of stars. Sweeping some 50,000 light-years across the disk of the Galaxy, this view reveals the blue-white stars of the spiral arms. These are extremely hot giant stars, much younger than the Sun. The orange-red glow of the nucleus comes from its red giant stars. Within the disk are deep red patches—nebulae where newborn stars are found.

The huge ball of stars is one of over 100 globular clusters that form a halo around the Galaxy's center. A typical cluster contains over 100,000 stars billions of years older than the Sun. Many are cool red giants. But there are hotter stars too, stars that shine with blue-white, yellow, and orange light. The small, fuzzy, orange-red globes are distant globular clusters.

About two million light-years away, to the right of the nearest globular cluster, we can see the Andromeda Galaxy, with its two companion galaxies. And to the upper right of Andromeda is another nearby galaxy, M33 in Triangulum. These are our backyard neighbors in a Universe that seems to be endless.

Everything in the Galaxy rotates about the nucleus—the gases, the stars, every bit

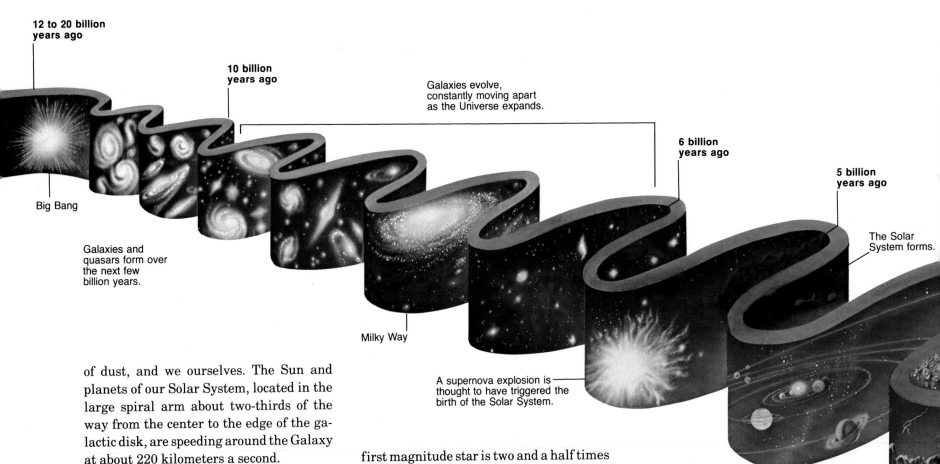

12 to 20 billion years ago

Big Bang

Galaxies and quasars form over the next few billion years.

10 billion years ago

Galaxies evolve, constantly moving apart as the Universe expands.

Milky Way

A supernova explosion is thought to have triggered the birth of the Solar System.

6 billion years ago

5 billion years ago

The Solar System forms.

Life begins on Earth: Amino acids, then proteins develop.

of dust, and we ourselves. The Sun and planets of our Solar System, located in the large spiral arm about two-thirds of the way from the center to the edge of the galactic disk, are speeding around the Galaxy at about 220 kilometers a second.

Fires in the sky

Greek scholars around 500 B.C. taught that the stars were flaming objects. Still earlier, people thought of meteors as "shooting stars" which fell to Earth. Today we know that the stars do not fall out of the sky and that they are not burning lumps of solid matter, but glowing balls of gases.

On any clear, dark night away from city lights, you can count about 2,000 stars. The measurement of a star's light as seen from Earth is its *apparent magnitude,* or *apparent brightness*. Since the time of Hipparchus, around 130 B.C., apparent magnitude has been a handy way of measuring the relative brightness of stars. Hipparchus and, later, Ptolemy rated stars on a descending scale from the brightest they could see (magnitude 1) to the faintest (magnitude 6). Our system is similar. A

first magnitude star is two and a half times brighter than a second magnitude star. A second magnitude star is two and a half times brighter than a third magnitude star, and so on. Adjustments to the magnitude scale and more precise measurements have made negative numbers necessary for the brighter-appearing stars. Sirius, for instance, is –1.5 and the Sun is –26.7. The faintest stars you can see with your unaided eye are of apparent magnitude 6.

With 7-power binoculars you can see over 50,000 stars, and with a 3-inch telescope you can see hundreds of thousands. Even though the Sun's bright light masks other stars from view during the day, they are there. As the sunlight fades in the evening, the stars blink into view, twinkling as their light is disturbed by the shimmering atmosphere. Have you ever wondered how far away those stars are? What makes them shine?

Orion the hunter

The constellation Orion, which dominates the winter sky from December through March, offers a wealth of interesting sky objects. It has more bright stars than any other constellation. Among them is the red supergiant Betelgeuse, of magnitude 0.4. It marks Orion's right shoulder. Betelgeuse is so large that if it were in the Sun's position, it would fill the Solar System out beyond the orbit of Mars.

Astronomers' figures for the sizes and distances of stars beyond the Sun are estimates, not precise measurements. Betelgeuse is estimated to be 520 *light-years*

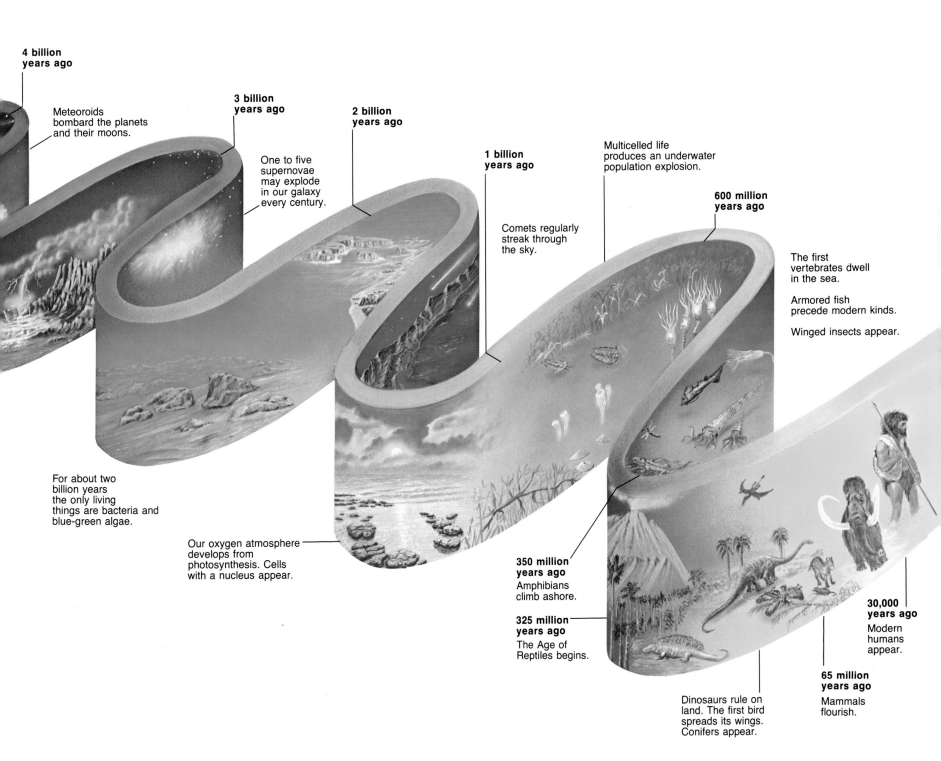

Some 12 to 20 billion years ago, astronomers think, a "primeval atom" exploded with a big bang sending the entire Universe flying out at incredible speeds. Eventually matter cooled and condensed into galaxies and stars. Planets formed around at least one star. Aeons after life began to develop on Earth, humans appeared. Here are major events in the history of the Universe. If all events until now were squeezed into 24 hours, Earth wouldn't form until late afternoon. Humans would have existed for only two seconds.

4 billion years ago

Meteoroids bombard the planets and their moons.

3 billion years ago

One to five supernovae may explode in our galaxy every century.

2 billion years ago

1 billion years ago

Comets regularly streak through the sky.

Multicelled life produces an underwater population explosion.

600 million years ago

The first vertebrates dwell in the sea.

Armored fish precede modern kinds.

Winged insects appear.

For about two billion years the only living things are bacteria and blue-green algae.

Our oxygen atmosphere develops from photosynthesis. Cells with a nucleus appear.

350 million years ago
Amphibians climb ashore.

325 million years ago
The Age of Reptiles begins.

Dinosaurs rule on land. The first bird spreads its wings. Conifers appear.

65 million years ago

Mammals flourish.

30,000 years ago
Modern humans appear.

away from us. (One light-year is the *distance* light travels in one year at nearly 300,000 kilometers a second. That means it moves at 186,000 miles a second or 669,600,000 miles an hour. This is the speed of light.) If the light reaching us tonight from Betelgeuse left that star 520 years ago, we are seeing Betelgeuse as it was about 18 years before Columbus set sail for America. If Betelgeuse exploded tonight, the flash would not be seen from Earth for 520 years. So when we look at the distant stars we look back in time. We can never see a star as it is in our "now" time.

Orion is also famous for its blue supergiant star Rigel, 900 light-years away and with a magnitude of 0.1. Rigel marks Orion's left leg and is the seventh brightest star in the sky. A small telescope shows that Rigel has a companion, Rigel B, which actually is a double star, or *binary*. Rigel B's two bluish-white stars revolve around each other. Bellatrix, 470 light-years away, marks Orion's left shoulder. The topmost of the three stars that form Orion's belt—1,500 light-years away—is Mintaka, another binary, of magnitude 2.2. The bottom star of the belt is a triple star, consisting of a blue-white binary with a faint, distant companion.

Three stars seem to form Orion's sword, but binoculars show that the middle "star" is actually an enormous cloud of hot glowing gas. Known as the Great Nebula, or the Orion Nebula, it contains a star group called the Trapezium. Just below Orion's belt is another nebula, the Horsehead. Both nebulae are 1,500 light-years from Earth. But the dark Horsehead can be seen

only by a time-exposure photograph which brings out its bright background.

The sky is a marvelous theater of wonders. There are stars that differ greatly in color, brightness, and size. There are single, double, triple, and multiple stars. There are dark clouds of gas and dust, and clouds hot enough to glow as brightly as beacons in the eternal night of space.

The constellations

Stargazers of very long ago invented the constellations. The earliest of these imaginary shapes probably included that great belt of 12 constellations we call the Zodiac

(shown on pages 16-17). From Earth the planets appear to glide through the constellations of the Zodiac.

By Ptolemy's time 48 constellations were listed, and by 1600 astronomers recognized about 60. Today the official number has grown to 88. In this celestial zoo of star groups are 19 land animals, 13 humans, 10 water creatures, nine birds, a couple of centaurs, one dragon, a unicorn, and a head of hair. Don't be disappointed if you can't make out the fanciful figures as they are shown in books. No one else can either. The constellations are beautifully false figures. Because the stars move in relation to one another, all the constellations are ever so slowly changing their shapes. Although some stars move along at more than 100 kilometers a second, they are so far away we cannot see them moving.

Grouping stars by luminosity

The constellations are not a scientific way of grouping the stars. A better way is to relate certain features of stars. For instance, a blue star is hotter than a red star. So there is a relationship between color and temperature. A blue star radiates more energy than a cooler red star the same size. So there also is a relationship between the rate at which stars pour out energy—called *luminosity*—and their size and surface temperature. This relationship is a useful way to classify stars.

Between 1911 and 1913 two astronomers—a Dane named Ejnar Hertzsprung and an American, Henry Norris Russell—had discovered how useful this relationship was and how much it could teach them

The Big Dipper's stars seem to be the same distance from us, but they're not. The diagram below shows (right) how we see them from Earth, and (center) their true distances from us. Red giant Dubhe looks farthest from blue Alkaid, but actually it's Alkaid's nearest neighbor.

A time diagram of the Big Dipper (bottom) shows that although the patterns of stars may seem permanent, they slowly change. Arrows in the center indicate the direction of real motion. These stars zip along at up to 20 km a second, but they are too far away for us to sense the motion.

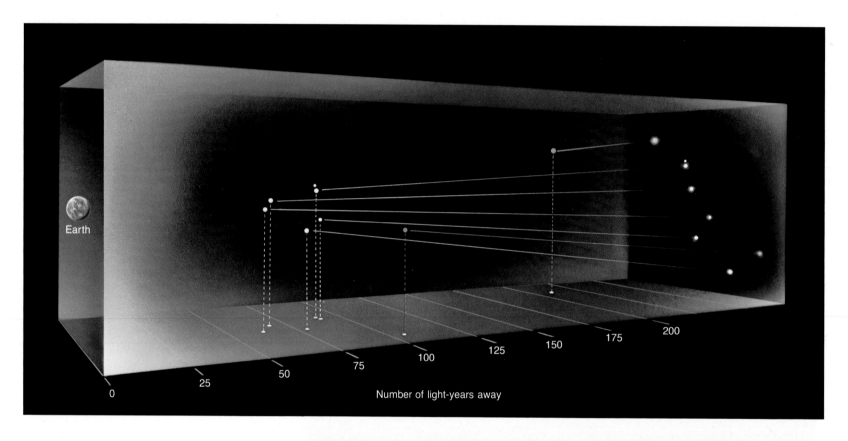

Earth

0   25   50   75   100   125   150   175   200

Number of light-years away

about stars. (See the graph on page 227.) Hertzsprung and Russell plotted the *absolute magnitude* against the *spectral class* and color of many stars. This type of graph was named in their honor the *Hertzsprung-Russell,* or *H-R, diagram.* It is also called the temperature-luminosity diagram.

When the two astronomers arranged stars on the diagram, they found that most fell along a narrow band from upper left to lower right—called the *main sequence.*

At the upper end of the main sequence are the massive and very luminous hot blue stars with surface temperatures up around 40,000 kelvins. Along the middle of

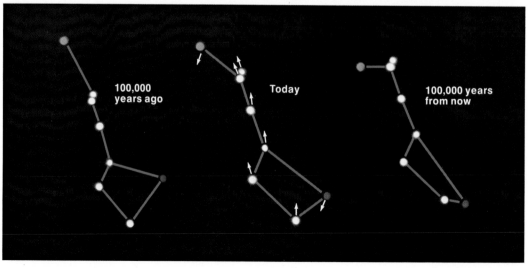

100,000 years ago

Today

100,000 years from now

the main sequence are less luminous and cooler stars, like the Sun. These stars shine with a yellowish-white light and have surface temperatures around 6,000 kelvins. At the lower end of the main sequence are the least luminous and still cooler red dwarf stars with surface temperatures of only about 3,000 kelvins. Although astronomers think that many very dim stars lie off the main sequence, about 95 percent of the stars we know today fall along the main sequence.

But where along the main sequence do most of the stars lie? When astronomers arranged the 100 stars nearest to Earth on a temperature-luminosity diagram they found that most fell along the lower end of the main sequence in the region of the red dwarfs. So most of the stars we can see are less luminous than the Sun. It turns out that the temperature-luminosity diagram is a very useful tool to help us study stars. The 100 brightest-appearing stars, for example, have been arranged on the diagram. Most of these are more luminous than the Sun but only a few are nearby. This reveals that highly luminous stars are rare among the nearest stars. Stars fainter than the Sun seem to be the rule.

Stellar mass and luminosity

Since stars differ so greatly in luminosity and size, do they also differ greatly in mass? In fact, the luminosity of a main sequence star depends on its mass. So here is another important relationship. The general rule is that the more luminous a main sequence star, the more mass it has. But it turns out that there is not nearly as much

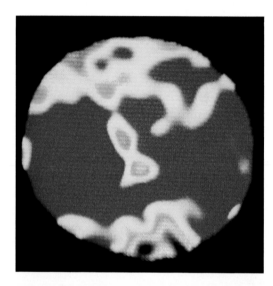

A computer portrait of Betelgeuse colors hot areas orange and cooler ones blue. These huge blotches may be convection cells like our Sun's granules (see page 49). Photographing Betelgeuse across 520 light-years is like photographing a sand grain a mile away.

difference in the mass of stars as there is in luminosity. Most of the stars that we can see range from about a tenth as massive as the Sun to about 10 times more massive. There are some exceptions where a star may be much more or much less massive. But we know of few such stars.

A star that is very massive shines and eventually goes out in a way very different from the way a red dwarf star with very little mass shines and ends its life. And a medium-mass star like the Sun has a life-style and fate different from those of a massive supergiant or a low-mass dwarf. The stars seem eternal and unchanging, but they are not. Stars are born, shine for a while, and eventually go out.

Because the life span of a star like the Sun stretches over billions of years, we cannot watch what happens to the star from the time it is formed until it goes out. But here and there in the sky we can see what appear to be stars forming. We also can study certain other stars in their youth, others in old age, and the remains of others that long ago blew themselves to bits in catastrophic explosions. So by splicing together these single scenes in the life of a certain type of star, we can make intelligent guesses about the evolution of a star

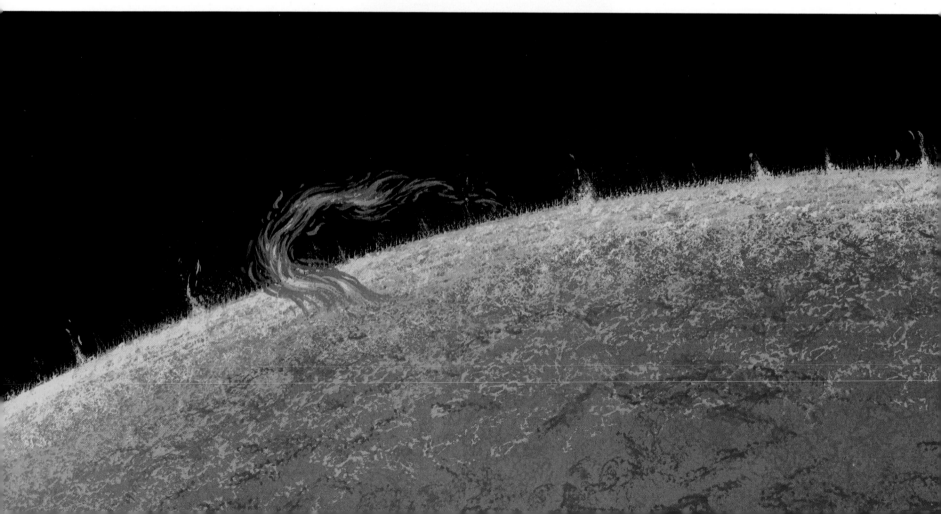

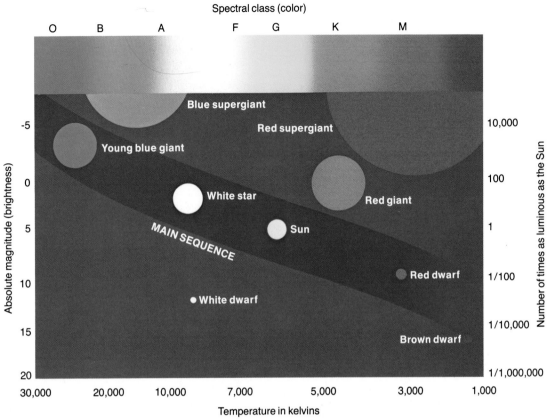

Spectral class (color)

O B A F G K M

Absolute magnitude (brightness)

Number of times as luminous as the Sun

-5 ... 10,000
0 ... 100
5 ... 1
10 ... 1/100
15 ... 1/10,000
20 ... 1/1,000,000

Temperature in kelvins

30,000 20,000 10,000 7,000 5,000 3,000 1,000

Blue supergiant

Red supergiant

Young blue giant

White star

Red giant

MAIN SEQUENCE

Sun

Red dwarf

White dwarf

Brown dwarf

This H-R diagram plots the actual brightness (absolute magnitude) of stars against their temperature, luminosity, and spectral class (the characteristics of the spectrum of each star). Absolute magnitude is the brightness stars would have if measured a standard distance away. A –5 magnitude star is very bright. Most stars fall along the main sequence. Surface temperature and size are related to brightness (but sizes at left aren't to scale): The hotter and bigger a star the brighter it glows. Red supergiant Betelgeuse is very bright, though cool, because its surface is huge—about 800 times bigger than our Sun.

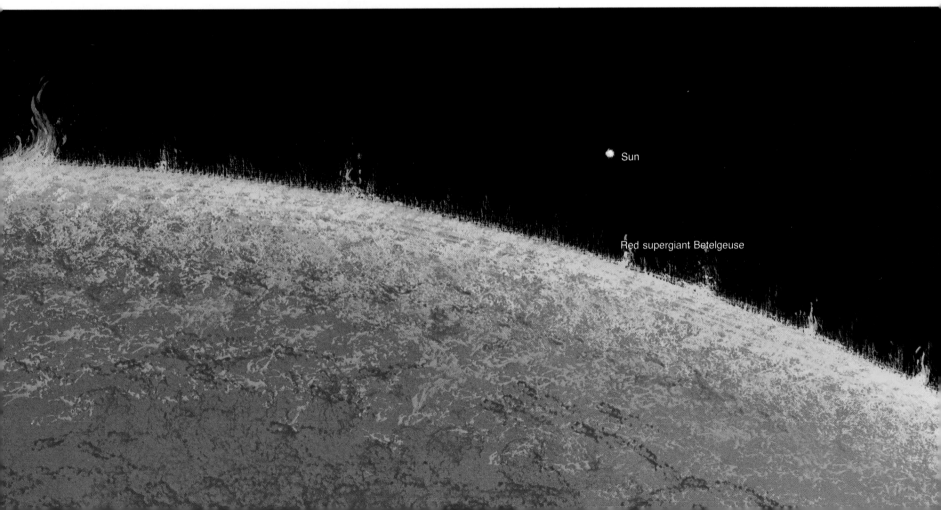

Sun

Red supergiant Betelgeuse

like the Sun, a blue supergiant like Rigel, or a red dwarf like Ross 128.

The birthplaces of stars

Astronomers think that stars are born in the *nebulae,* those clouds of gas and dust that we see in the Milky Way and that exist in many other galaxies. The gas is mostly hydrogen with some helium. Space dust is made up of tiny particles of solid matter.

Some are *dark nebulae.* These are clouds sometimes silhouetted by the light of stars or gas shining behind them, such as the Horsehead Nebula. The large dark band, or *rift,* visible along the summer Milky Way is a series of dark nebulae. When a cloud of gas and dust has one or more stars inside or nearby, the nebula may shine by reflecting the starlight. These are called *reflection nebulae.* The Pleiades star cluster (page 233) contains several well-known reflection nebulae.

If the star or stars inside a nebula heat the gas to about 10,000 kelvins by ultraviolet radiation, the hydrogen of the nebula is excited and glows with a fluorescent light. Nebulae that emit light in this way are called *emission nebulae.* The most famous one is the Great Nebula in Orion.

How stars are born

Stars form out of especially dense clouds of gas and dust. A globe of this matter grows as it keeps sweeping up surrounding matter by gravitational attraction. As such a *protostar* globe continues to gain mass, gravitation packs the matter tighter and tighter in the core region, where the pressure, density, and temperature steadily

Double stars that orbit one another in our line of sight can block each other's light. The binary Algol dips its light every 69 hours as its dim star obscures its bright one (top). Known as the "winking demon," Algol's brightness dips again, but less so, when the dim star is eclipsed.

Mira, Mira, burning bright, growing larger dims your light. Like the red giant Mira "the miraculous" (bottom), some stars change size periodically, varying their brightness. Over an 11-month cycle, Mira expands and contracts. It's brightest when small, dimmest when large.

climb. The core of the protostar heats up enough so the star glows a dull red. As it becomes still hotter it glows a brighter red. Eventually, when the core temperature reaches about 10 million kelvins, nuclear fusions of hydrogen into helium begin. (Page 61 explains fusion.) Soon after its nuclear furnace ignites, the new star becomes a main sequence star. An object that forms like a protostar, but which is too small to ignite hydrogen fusion, is called a *brown dwarf*. Astronomers have found several objects that may be brown dwarfs.

During the time a protostar is pulling in matter from its parent nebula, it is spinning faster and faster. A large disk of revolving matter forms around it. Small globes of matter may form out of a star's disk material and become *protoplanets*. These do not have enough mass to heat up and become stars. Instead they condense and become planets. If planets often form around new stars, there might be billions of other planets in our galaxy alone. Vega, 26 light-years away, is one of several stars that may have planets forming.

During the protostar stage of star formation, a planetary system must be a foggy place, glowing dull red with the feeble light of its new star. Eventually, however, the star may go through a stage in which it pours out gas particles in the form of a powerful wind. If so, the wind, called the T Tauri wind, will sweep the planetary system clean of gas and dust. According to this theory, if the Sun passed through a T Tauri stage, it is possible that the strong wind stripped Mercury, Venus, Earth, and Mars of their primitive hydrogen and helium

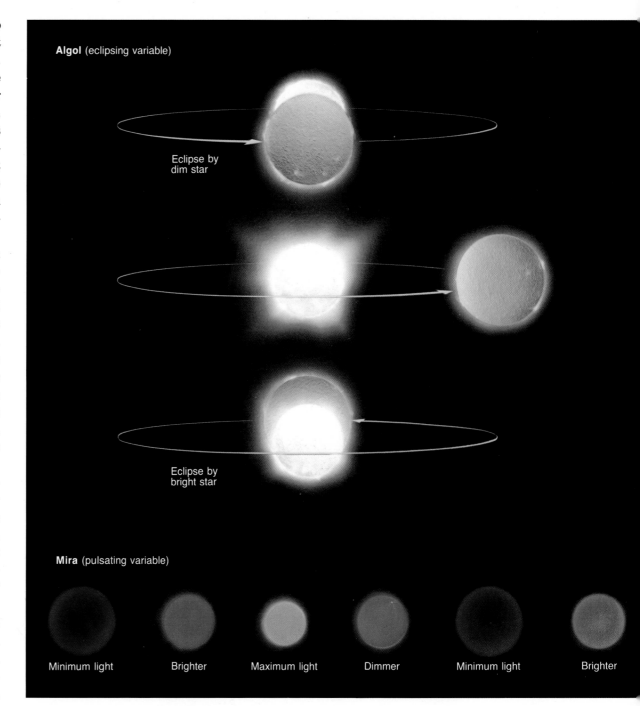

Algol (eclipsing variable)

Eclipse by dim star

Eclipse by bright star

Mira (pulsating variable)

Minimum light | Brighter | Maximum light | Dimmer | Minimum light | Brighter

1 Nebula and protostar

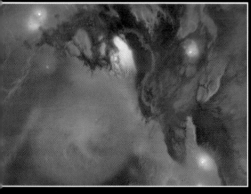

2 Star and planets forming

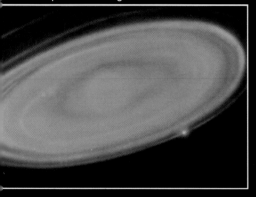

3 Star on the main sequence

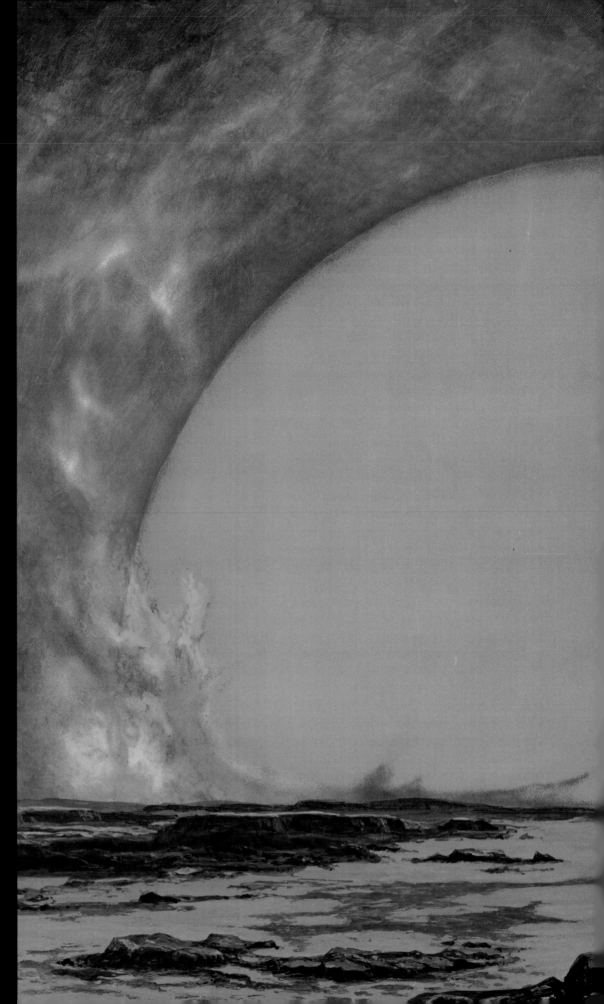

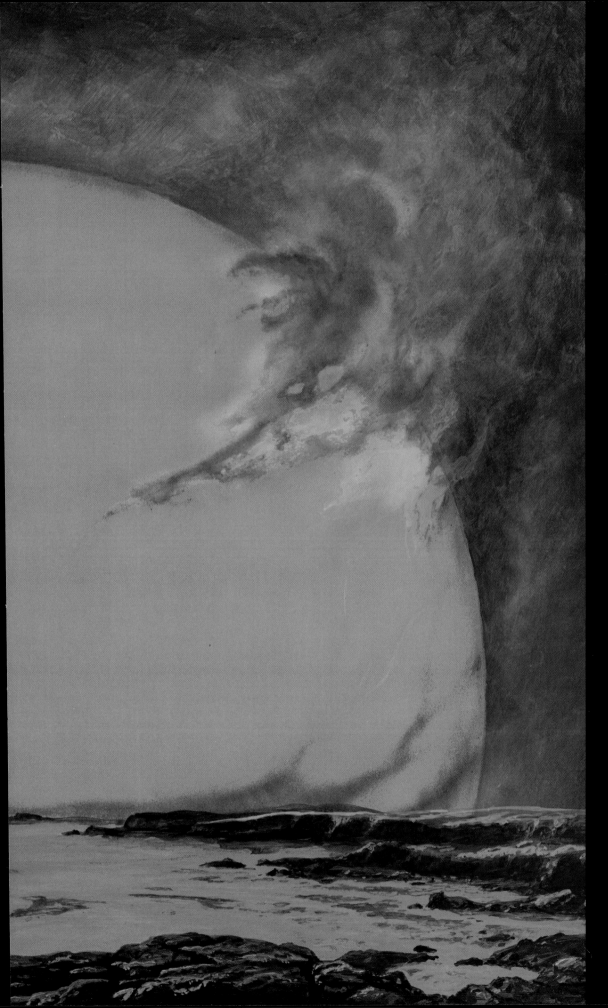

4 Looming red giant

5 Planetary nebula and central star

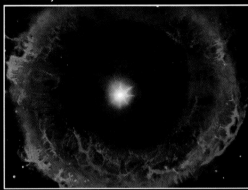

6 White dwarf seen from a cold planet

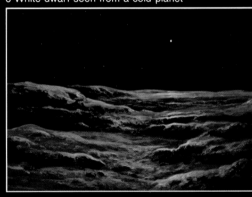

(1) A star like our Sun is born inside a huge cloud of gas and dust; gravity condenses part of the cloud into a warm, red protostar. (2) Spinning flattens it into a disk, the infant star at its core; planets may form from rings in the disk. (3) Contraction heats the core until the start of nuclear reactions that make the star shine for billions of years. (4) Its hydrogen fuel spent, the star's core contracts further, yet radiation swells the outer layers to an enormous size. (5) Later, these layers blow off in an expanding cloud of gas and (6) leave the star's burnt-out core to cool and fade.

A dark dust cloud, the Horsehead Nebula, takes shape against a field of luminous gas in the constellation Orion. Like the hundreds of other nebulae in our galaxy, the Horsehead is a birthplace of stars. Such dark nebulae resemble bright ones but lack the stars to make them shine.

The Trifid Nebula (below right) glows red as ultraviolet rays from its hot central star excite its hydrogen gas. Because the Trifid emits visible light by a fluorescent process, it is called an "emission" nebula. Dark dust alleys between Earth and the nebula divide its bright regions into three parts.

Bright halos of cloud surround the stars of the Pleiades, or Seven Sisters cluster. These nebulae glow by reflecting light from stars. Dust grains in the halos scatter the blue-white light of the nebulae's young stars the way Earth's atmosphere scatters sunlight and makes the sky blue.

atmospheres. Jupiter and the other present-day gas giants managed to hold on to their gases, perhaps because of their greater distances from the Sun.

Once a protostar becomes a main sequence star, it shines more or less steadily for millions or billions of years. But just how stable an adult star remains, and just how long it stays a main sequence star, depends on how much mass the star has.

The red dwarfs are the least massive stars we know of. Because of their low mass, their core temperatures reach only about 10 million kelvins, just hot enough to start hydrogen fusion reactions. So these stars shine with a cool red light and have surface temperatures of only about 3,000 kelvins. The red dwarfs burn up their hydrogen fuel very slowly so they have very long lives, perhaps a trillion years or so.

Stars like the Sun, about 10 times more massive than the red dwarfs, have core temperatures around 15 million kelvins. They radiate energy much more rapidly and so have surface temperatures around 6,000 kelvins. Since Sunlike stars use up their hydrogen fuel more rapidly than the red dwarfs do, they have shorter lives, around 10 billion or so years.

Very massive stars, like the Orion belt stars, are so hot that they shine with a bluish-white light. Some of these stars have surface temperatures around 30,000 kelvins or more. Their enormous mass of about 35 times that of the Sun drives their core temperatures up to 40 million kelvins. These stars are energy spendthrifts and use up their fuel supply rapidly. Their lives span only millions of years.

How stars die

The smallest red dwarfs evolve so slowly that even one formed in the earliest days of the Universe is burning unchanged on the main sequence and will stay there for the entire lifetime of the Universe. But a star that does leave the main sequence, no matter how much or how little mass it has, eventually swells up into a red giant before burning out. The star's mass determines how dramatically it ends its days. Picture a low-mass star down in the red dwarf region of the main sequence nearing the end of its hydrogen fuel supply. Gradually it sends less and less energy out from the core region and so begins to cool. A cooling core is a core with decreasing pressure, so the star collapses in on itself.

This sudden gravitational fall of matter into the core temporarily sends the temperature and pressure zooming. The star swells up and becomes a red giant for a while. Many stars that pass through the red giant stage cast off one or more shells of hydrogen and other gases. These shells balloon out and become *planetary nebulae.* (In spite of their name, they have nothing to do with planets.) The famous Ring Nebula in the constellation Lyra is a splendid planetary nebula. These gas shells cast off by a dying star glow with fluorescent light for 100,000 years before they fade.

A lower main sequence dwarf star passes through the red giant stage and then slowly cools and contracts. Its nuclear furnace has been turned off for good. It now radiates heat and shines by the infall of its matter packing itself tighter in the core. Eventually the star shrinks to about the

Like a celestial lighthouse, a spinning neutron star—called a pulsar—emits a strong radio beam that we detect as a pulse as it sweeps past Earth. When radio telescopes first picked up pulsar bursts in 1967, scientists thought they could be messages from space and called them LGM, "little green men." A pulsar's energy may come from a "hot spot" near its surface, at or above a magnetic pole. Some pulsars send out X rays, and two emit light waves also. Neutron stars are so dense, a teaspoonful of one would weigh a billion tons on Earth.

Pulsar

Magnetic pole

Rotating radio beam

Radio signal received on Earth

size of Earth. It now has so little surface area from which to send out its energy that it shines with a faint white light. It has become a *white dwarf*. Over billions of years it continues to cool, gradually dims, and goes out, a *black dwarf*.

Higher up along the main sequence are medium-mass stars like the Sun. When such a star burns up its hydrogen fuel and collapses in on itself, it also swells up as a red giant. The core temperature zooms much higher than in a low-mass dwarf star, high enough so that the helium is fused into carbon and other elements. This burst of heat also causes a shell of hydrogen just outside the core to fuse into helium, so the red giant's nuclear furnace is still blazing away. Because the star lacks enough mass to keep the temperature high, the fusions stop and these medium-mass stars also collapse into white dwarfs and finally go out. But like less massive dwarfs, these medium-mass stars also may go through a planetary nebula stage.

Extremely massive stars—30 or more times more massive than the Sun—keep the fusion process going until their cores are almost pure iron. Wrapped around the iron core is a hot shell of silicon fusing into iron, then a less hot shell of carbon and oxygen fusing into silicon, then a still less hot shell of helium fusing into carbon and oxygen, then a cooler shell of hydrogen fusing into helium, and finally an outer atmosphere of hydrogen. At this stage the very massive dying star is a supergiant large enough to fill much of our Solar System.

As far as we know, elements heavier than iron—gold, lead, silver—cannot be

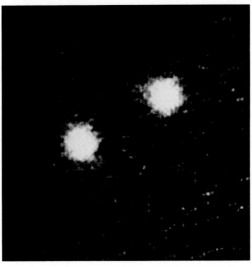

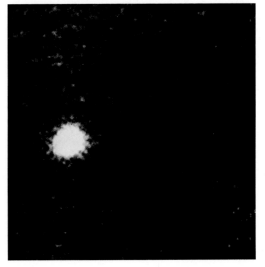

Can you imagine a star so massive that its gravitation eventually crushes it out of existence, leaving only a black hole in the sky? Many scientists can. In this artist's dramatization (opposite), a black hole's powerful gravitational field sucks in a light ray. A red star is beyond its reach.

What goes in never escapes a black hole. That includes light and other energy, so scientists can't observe one directly. But they have found suspects. One, called Cygnus X-1 (below), seems to pull gases off another star to form a rotating disk that heats up and emits detectable X rays.

fused in any star's core. Then where do the heavier elements of the Universe come from? A high-mass star with an iron core heated to a billion kelvins collapses, causing a catastrophic explosion known as a *supernova*. The Crab Nebula is the remnant of a supernova seen to explode in the year 1054. Such explosions are thought to produce all the known elements heavier than iron. Dying stars, then, are the element factories of the Universe.

Neutron stars

A supernova may leave an extremely dense core—a tightly packed ball of subatomic particles called neutrons. Because neutrons, unlike protons, do not have an electrical charge, they do not repel each other. Therefore they can be packed together extremely tightly. These dense cores have masses of about one and a half times that of the Sun. They survive the supernova as superdense neutron stars, only 20 kilometers across.

Radio astronomers first detected neutron stars in 1967. Some exist in binary systems that release huge amounts of X rays. Others emit a continuous beam of radiation as they rapidly spin. Each time the beam crosses Earth, astronomers can detect a radio "pulse." These rotating neutron stars are called *pulsars*. The central bright object in the Crab Nebula is now known to be a pulsar rotating about 30 times a second. All pulsars slow down and are destined eventually to fade away.

Black holes—the end?

Stars even more massive than those that

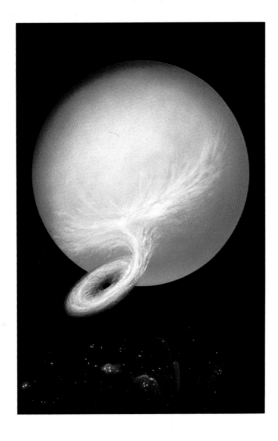

become neutron stars have a strange fate in store—that of a *black hole*. One astronomer describes a black hole as an object which dug a hole, jumped in, and then pulled the hole in after itself! Some astronomers doubt that these mystery objects exist.

Those who do believe in black holes describe them this way. When a burned-out star, say one 10 times more massive than the Sun, shrinks to about 60 kilometers in diameter it becomes very special. It is so dense, and its gravity so strong, that virtually nothing can escape from it, not even light. At this great density the star disap-

pears inside itself, or within its so-called *event horizon*. After a dying star contracts into a black hole, it continues to contract until it is a point called a *singularity*.

If energy cannot escape from a black hole, how do astronomers detect one? Gaseous matter swirling toward a black hole from a visible companion star is heated to about a billion kelvins. The hot gas glows in X rays. The visible star orbits as though it were a member of a binary system; its partner must be an unseen black hole. The X-ray source known as LMC X-3 in the Large Magellanic Cloud is one of the best candidates for such a binary system. Astronomers have increasing evidence that the centers of galaxies contain black holes millions of times more massive than the Sun.

Variable stars

Many stars go through cycles of brightening and dimming. They are then called *variable stars*. Over 25,000 have been catalogued, but most stars may become variables at some stage. They fall into three main groups: explosive, such as novae and supernovae; eclipsing, such as the Algol binary (page 229); and pulsating variables, which rhythmically swell up and shrink. What makes the latter go from bright to dim and then back to bright again in regular *periods* is not well understood. A few Mira-type variables, such as Mira itself (page 229), are visible to the naked eye, but most are too faint to be seen without a telescope. These are red giant or red supergiant stars with surface temperatures only about 2,000 kelvins and periods that range

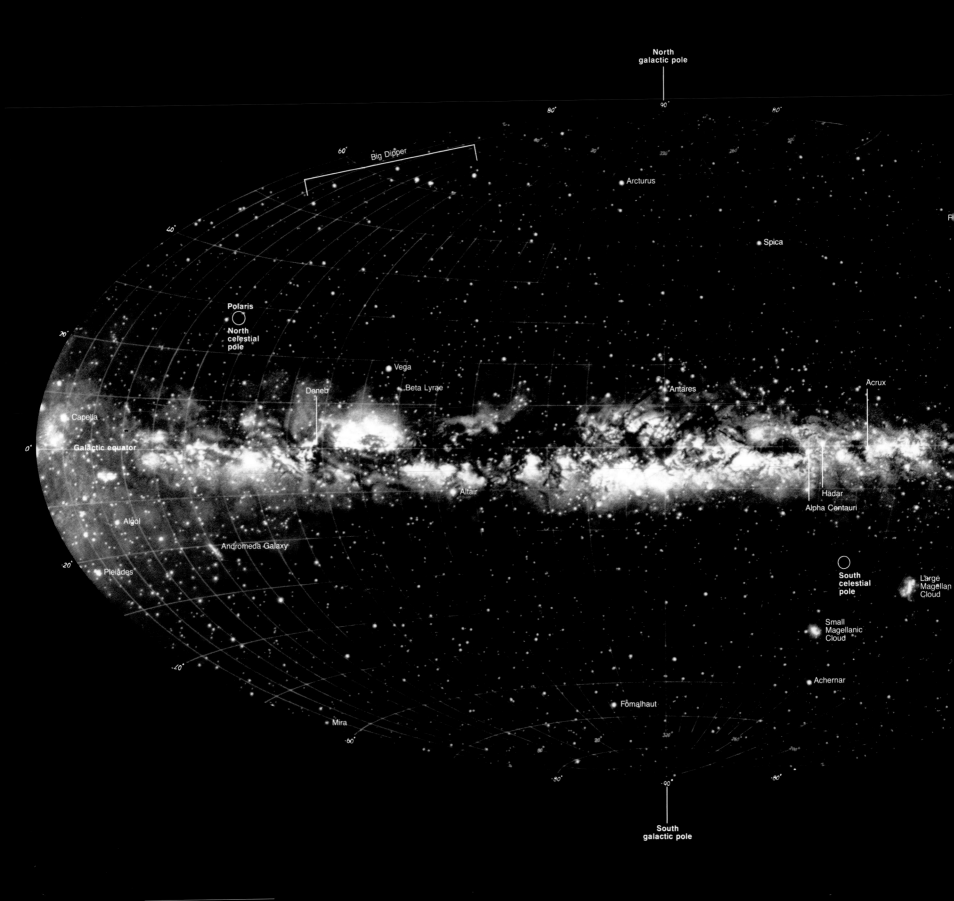

North
galactic pole

Big Dipper

Arcturus

Spica

Acrux

Polaris

North
celestial
pole

Vega

Beta Lyrae

Antares

Capella

Deneb

Galactic equator

Altair

Hadar

Alpha Centauri

Algol

Andromeda Galaxy

South
celestial
pole

Large
Magellan
Cloud

Pleiades

Small
Magellanic
Cloud

Achernar

Fomalhaut

Mira

South
galactic pole

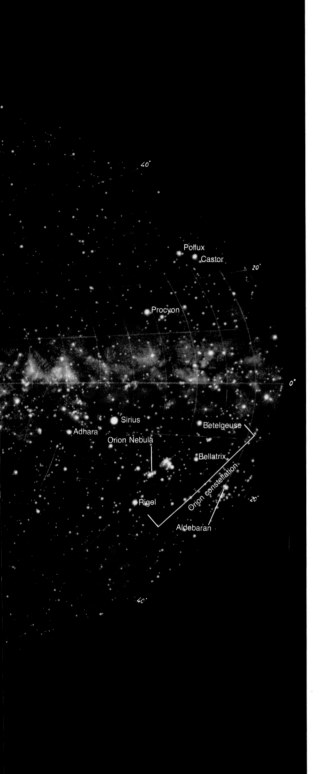

Because we're inside the Milky Way Galaxy (see page 243), what we see of the spiral arms around us is a cloudy band, also called the Milky Way, that stretches across the night sky. Think of a transparent sphere around Earth with all the visible stars fastened inside. If you slit the sphere and stretched it back (see below) you'd have a map something like this (left), with constellation shapes distorted at the edges of the projection. The equator extends along the plane of the Milky Way Galaxy. The Galaxy's center lies hidden, about 23,000 light-years away.

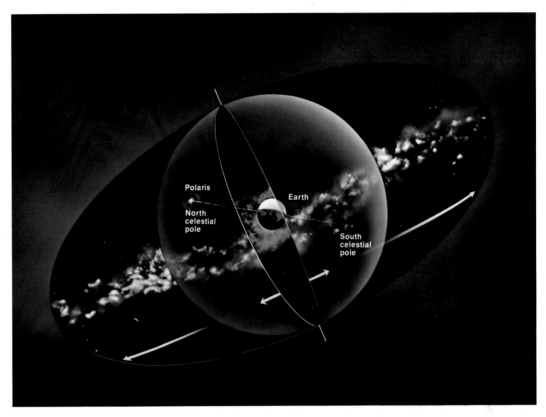

from 80 to 1,000 days. When bright, a Mira variable emits about three times more energy than when dim, but its apparent brightness increases about 200 times.

We know of about 5,700 variables called RR Lyrae variables. These are white or yellow-white giant stars with short periods of less than a day. They all seem to have the same luminosity, thus the apparent brightness of such a star tells us how far away it is. The center of our galaxy and the globular clusters contain RR Lyrae variables. Since astronomers can find the distance to these variables, they can estimate the distance to the globular clusters.

The 900 or so Cepheids are the best known variables. They are huge, highly luminous stars with periods of a few days to about 50 days. These stars are so bright that astronomers use them as celestial "yardsticks" to estimate the distance to galaxies beyond our own. After a few million years of pulsating, a variable star may resume shining at a steady rate.

Exploring our galaxy

We live in a vast swirling city of stars called a *galaxy*. Our home galaxy is called the Milky Way, or the Galaxy, and is much like the Andromeda Galaxy (page 240).

More luminous than 20 billion Suns, Andromeda Galaxy (opposite) looks from Earth like a thin cloud twice the size of our Moon. Astronomers used to wonder about this and other fuzzy patches in the sky. Were they "island universes" of their own? In 1924 Edwin Hubble proved

they are other galaxies far beyond ours. He classed them as spiral, barred spiral, elliptical, irregular. Like our galaxy, Andromeda is a spiral; its two companions are elliptical. NGC 5128 is "peculiar" because of its dust lane. The Large Magellanic Cloud is nearest to Earth.

The Milky Way is a discus-shaped collection of about 200 billion stars. It stretches some 70,000 light-years from edge to edge and is over 1,000 light-years thick. In addition to its stars, there are huge amounts of interstellar gas and dust and there may be untold billions of planets with their moons. At the center is the *galactic nucleus,* surrounded by a huge central bulge of stars about 10,000 light-years across. Around the fat galactic center is the thin *galactic disk* with stars of every description and age: stars now being born and stars ending their lives; open clusters of 100 to 1,000 stars—such as the Pleiades—born out of a single nebula; and a seemingly endless number of nebulae.

All of these objects are arranged in a pinwheel pattern of spiral arms. All revolve around the galactic nucleus, much as the planets revolve about the Sun. So Newton's law of gravitation works just as well for a galaxy as it does for an apple or a planetary system. Speeding along at some 220 kilometers a second, the Sun takes about 230 million years to circle the Galaxy once. This is a *galactic year.* Optical telescopes cannot see through the dense nebulae to the center of the Galaxy and beyond. But radio telescopes can penetrate the murky soup and map the distant corners of the Milky Way.

The spiral arms contain the blue giant stars and clouds of gas and dust—the birthplace of stars. Between the arms are the older Sunlike stars. Our Sun is out in the galactic suburbs some 23,000 light-years from the center. Our galaxy has a spherical halo of globular clusters, hundreds of

NGC 2997 Spiral

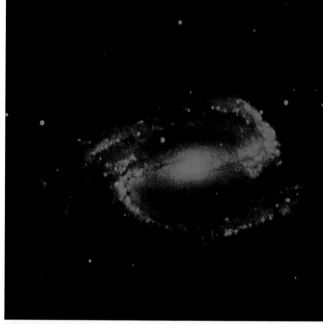
NGC 1300 Barred spiral

NGC 5128 Elliptical (peculiar)

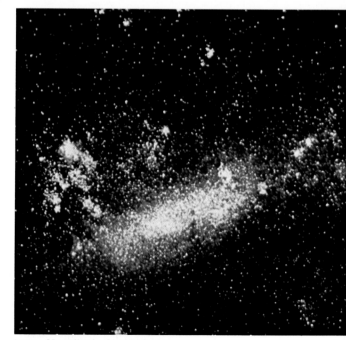
Large Magellanic Cloud Irregular

One corner of the Universe

So vast is space, so insignificant our tiny corner of it, that we need a four-stage view – four leaps of scale – just to find the Solar System. The Universe extends from Earth for at least 10 billion light-years. In the background on these two pages, you see a mere sliver of the Universe, yet even the smallest dots are not stars, but entire galaxies. Think of seeing this scene in an imaginary telescope. Turn up the power and search. Our Local Group of galaxies pops into view (below). More magnification, and the Milky Way alone fills the eyepiece (right). But our Solar System is still far too small to spot. To see it, we must imagine the biggest jump of all. Zoom in on a shining spiral arm, the Orion Arm. Go closer; the swirl of light sharpens into thick clouds of bright points – stars, like powder spilled on black velvet. Still closer; the stars seem

Local
Group of
galaxies

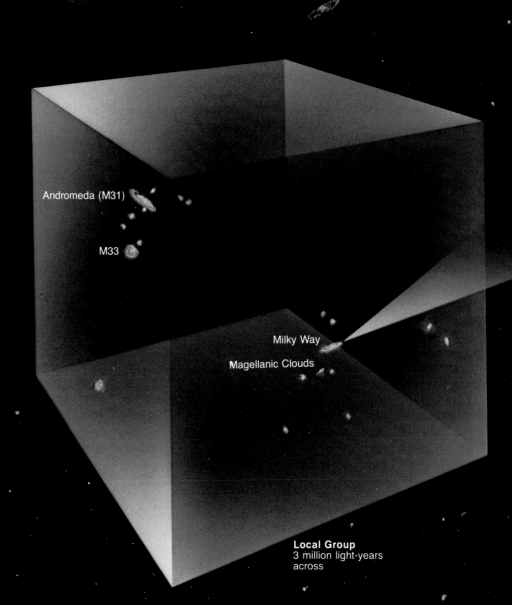

Andromeda (M31)

M33

Milky Way

Magellanic Clouds

Local Group
3 million light-years
across

242

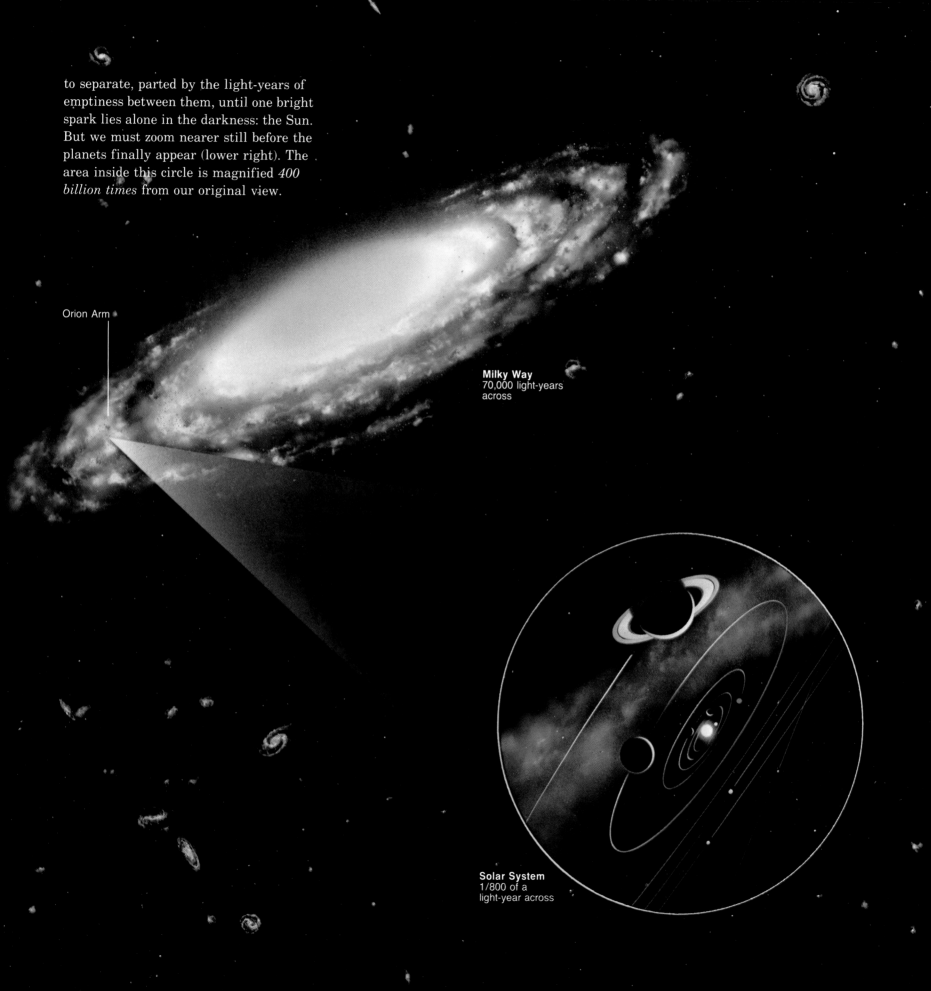

to separate, parted by the light-years of emptiness between them, until one bright spark lies alone in the darkness: the Sun. But we must zoom nearer still before the planets finally appear (lower right). The area inside this circle is magnified *400 billion times* from our original view.

Orion Arm

Milky Way
70,000 light-years
across

Solar System
1/800 of a
light-year across

The spaceships (opposite) suggest what happens to the spectral lines of a star or galaxy zooming away from us or moving toward us: they shift to the red or blue end of the spectrum (see diagrams below). That's because of the Doppler effect — a change in light or sound wavelengths when their source moves in relation to the observer. The light waves of an object hurtling away through space will stretch into longer wavelengths, or "red-shift". In this color-coded radio map of a rotating galaxy (top picture) blue areas advance while red ones recede.

thousands to millions of stars. We know of about 125 clusters. They contain the oldest known stars in the Galaxy. Measurements of their age suggest that the Galaxy formed about 15 billion years ago.

Galaxies galore

In the late 1700's William Herschel said that certain fuzzy patches which looked like nebulae might actually be galaxies. In 1924 the American astronomer Edwin Hubble, using the 100-inch Mount Wilson telescope, clearly showed that the so-called Andromeda "nebula" was actually a spiral galaxy far beyond our own. By the study of Cepheid variables, we know that Andromeda is about two million light-years away.

As stars differ from one another, so do galaxies. There are *spiral galaxies* like Andromeda and the Milky Way. While some have tightly wrapped arms, in others the arms are loosely wrapped. Spiral galaxies have reddish-orange centers because of the many old red giant stars located there. Red dwarfs make up about 90 percent of the stars in the galaxies, but are very faint. The young bright stars out in the disk contribute most of the light. Some spirals have bars running through their centers (like the barred spiral galaxy on page 241). There also are *elliptical galaxies,* some shaped like a football, others like a sphere. Elliptical galaxies also seem to contain mostly old stars.

Irregular galaxies, such as the Large Magellanic Cloud, have no special shape. *Peculiar galaxies* have some rare or particular feature — an exploding central region, for instance. Supergiant elliptical galaxies

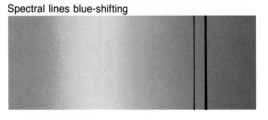

Spectral lines blue-shifting

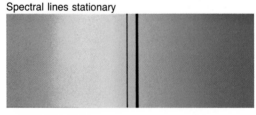

Spectral lines stationary

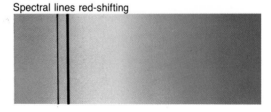

Spectral lines red-shifting

are the largest known, while some dwarf ellipticals are not much larger than a globular cluster. The dwarf ellipticals greatly outnumber the supergiant ellipticals, just as red dwarf stars greatly outnumber supergiant stars.

Most galaxies are arranged in large groups called *clusters.* The Andromeda Galaxy has two elliptical companions (see page 240) and the Milky Way has two irregular companions, the Large and Small Magellanic Clouds, visible only from the Southern Hemisphere. Our *Local Group* cluster of galaxies contains about 21 galaxies, more than half of which are ellipticals. The Local Group has a diameter of about three million light-years. Clusters of galaxies stretch off into space for as far as we can see with telescopes. Within 50 million light-years of us are many dozens similar in size to our Local Group.

Clusters of galaxies form superclusters. One supercluster in the constellation Hercules is thought to be 350 million light-years across. Possibly the system of clusters and superclusters continues into super-superclusters. So far as we can tell now, our lumpy Universe of galaxy clusters stretches off in every direction for at least 10 billion light-years.

The runaway Universe

No matter which direction we look in space we see the same thing — countless billions of galaxies all rushing away from us. A person in any one of those other galaxies would see the same rushing-away motion. Around 1930 Hubble discovered that the farther away a galaxy is from Earth, the

Long "tails" give this pair of colliding galaxies a nickname: the Mice. Mutual gravitational forces pulled out tails and formed a bridge of stars, gas, and dust. A computer sequence (below) shows what can happen when two galaxies like these get close: Gravitation pulls their disks apart.

faster it is speeding away. For example, the Coma Cluster, 291 million light-years from us, is speeding away at 6,700 kilometers a second; the Leo Cluster, 847 million light-years distant, has a speed of 19,500 kilometers a second; and the Hydra Cluster, 2.6 billion light-years away, has a speed of 60,600 kilometers a second.

The real speed demons are the *quasars,* the most luminous objects known. These mysterious points of light in the cores of galaxies are the most distant things we can see and therefore have the highest redshift velocity. Some travel over 90 percent the speed of light—more than 270,000 kilometers a second. In one second a quasar releases the energy to supply all of Earth's present electrical needs for a billion years.

So we live in a Universe that seems to be expanding at breakneck speed. What evidence do we have that the galaxies actually are rushing away from us? Just as astronomers photograph stars with a spectrograph to obtain the star's spectrum, they also photograph the spectrum of a galaxy. Such spectra show two dark lines produced by the element calcium. These dark lines are important bits of information. The light waves of an object speeding toward us get bunched up and its spectral lines shifted toward the violet end of the spectrum, away from their normal position. But the light waves of a galaxy speeding away from us get stretched out and are shifted down toward the red end of the spectrum. And that is just what happens to the dark calcium lines in the spectra of galaxies. The faster the runaway speed of the galaxy, the greater the red shift. For really

distant objects, like quasars, other lines must be used. But all lines are shifted toward the red end of the spectrum.

The Big Bang

Has the Universe always been expanding? Will it go on expanding forever? Many astronomers favor the Big Bang explanation of the expansion of the Universe, first proposed in 1927 by Belgian cosmologist Georges Lemaître, although some astronomers also think that another kind of Universe might have existed earlier. According to the Big Bang theory, the entire Universe began as an incredibly dense "primeval atom." This "atom" exploded 12 to 20 billion years ago with tremendous force, and all matter and space began expanding at speeds nearly that of light.

Between 100,000 and a million years after the Big Bang, enormous clouds of hydrogen and helium began to form. These were the only two elements present in abundance then. The clouds of gas became the protogalaxies. Eventually star formation began and, during the first few billion years, all the galaxies formed.

The first stars to form in those early galaxies were protostars of high mass, stars that became supernovae. On exploding, the supernovae enriched surrounding nebulae with heavy elements. Later generations of stars, like the Sun, formed out of these enriched nebulae and so contain a wide variety of elements.

Sometimes galaxies "collide." When they do they simply pass through each other, so great is the distance between individual stars. During such a sweep-through,

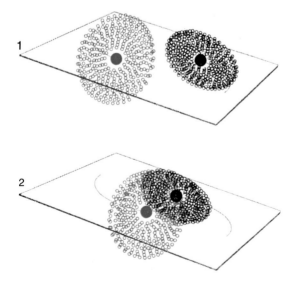

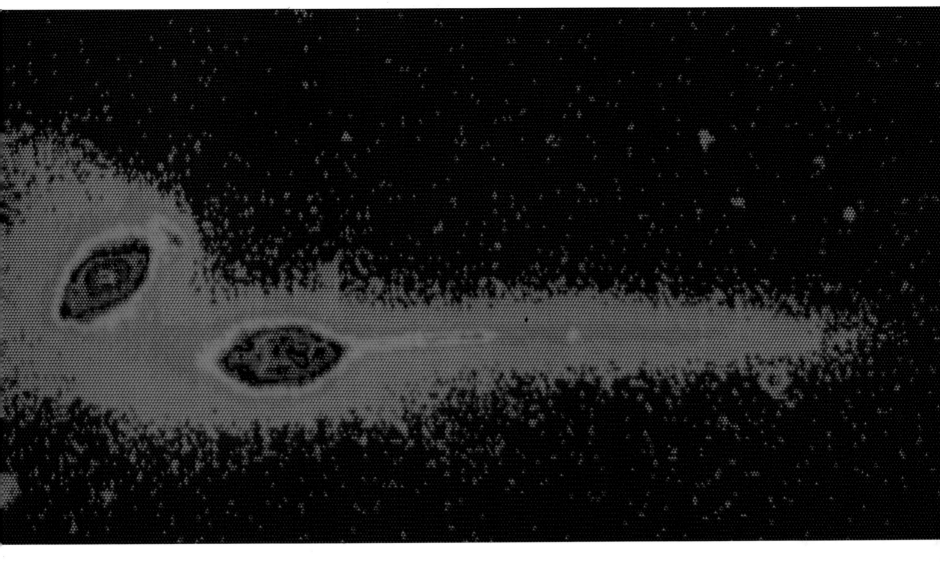

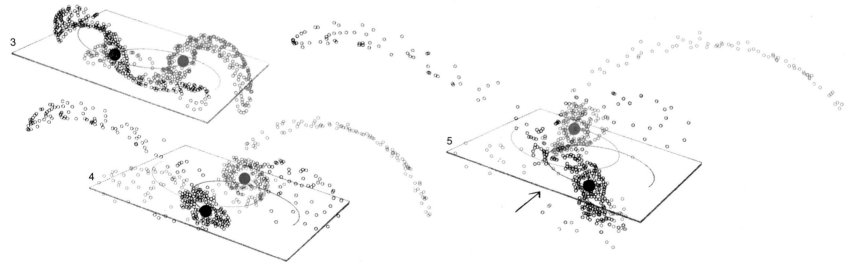

3

4

5

Another planetary system? One just beginning to form? In 1984 astronomers at an observatory in Chile used a 100-inch telescope and state-of-the-art imaging techniques to photograph what may be a new solar system. This massive cloud-like disk of dust surrounds the star Beta Pictoris, some 50 light-years from Earth. The disk extends nearly 60 billion kilometers to either side of the star, which is about twice the Sun's mass and about 10 times brighter. The shape of the dust cloud suggests to astronomers that planets have already begun to form.

the less massive galaxy may be cleaned of its gas and dust. Star formation can no longer take place there.

First the short-lived blue giants will flicker out, then the longer-lived yellow-white stars. Eventually the galaxy will glow dull red as the sole remaining stars, long-lived red dwarfs, themselves burn down and eventually leave the galaxy a dark, cold desolate place.

Will the Universe keep on expanding forever? Will something stop it? If the Universe is dense enough, gravitational attraction will slow down the expansion and reverse it, just as Earth's gravity slows down and reverses a handful of marbles tossed high into the air. Like the marbles, eventually the galaxies would slow down, stop and hang motionless for an instant, and then begin falling inward. Astronomers of the future would then see blue-shifted galaxies.

The Big Squeeze

If this happens all matter will tumble together again in the Big Squeeze billions of years from now. The Big Squeeze may form another cosmic egg that will explode in another Big Bang and start the process all over again. This is the oscillating Universe theory. Some astronomers call it the Bang-Bang-Bang theory.

One trouble with this idea of a Universe forming and destroying itself in cycles is that the Universe may not be dense enough to stop the present expansion. Perhaps we live in a Universe that will expand away forever and simply run down, never to be reborn.

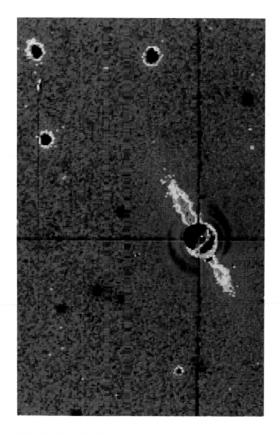

Life in the Universe

One of the most exciting questions we can ask is whether life exists on worlds other than our own. If so, what is it like?

Scientists looking into the origin of life on Earth have come to think that life may abound in our galaxy, and in other galaxies. They have found evidence that bacteria and simple plantlike organisms have existed on our planet for at least 3.5 billion years. Biologists think these organisms evolved from simpler nonliving matter. Then for about two billion years the plantlike organisms themselves evolved. Eventually they gave rise to the millions

of plants and animals we find in the fossil record down through time.

What conditions seem necessary for intelligent life to develop on a planet? First, a planet must not be too near or too far from its local star. If too near it will be too hot for certain life-giving molecules to form. And the liquid water needed for life-producing reactions would boil away. If too far it will be too cold for liquid water. Another important thing is the planet's mass. Too much mass may lead to harsh conditions such as the dense atmospheres on Jupiter and Saturn, poisonous to the life we know. Too little mass may lead to different, but equally harsh, conditions such as the lack of atmosphere and the extreme temperatures on Mercury and the Moon.

The kind of local star a planet has also is important. If it is a blue giant with a life span of only a few million years, the star will end its life before intelligent beings can evolve. If we use Earth as a model, a planet needs about four billion years to develop a technological civilization.

In spite of these needs, some astronomers estimate that in our galaxy alone there are hundreds of millions of planets able to support a technological civilization. They also think there are now about a million civilizations at or beyond Earth's level. Other astronomers doubt this.

The Vega discovery

In 1983 the Infrared Astronomical Satellite (IRAS) made an exciting discovery: a huge ring of solid particles around the star Vega that may be a planetary system forming. Although astronomers think the particles consist of material left over from Vega's formation about a billion years ago, they cannot yet decipher whether this material represents a new planetary system, or one that has failed to form. IRAS's limited resolution prevented the astronomers from making out the exact shape of the ring. But they know its diameter is twice the size of Pluto's orbit around the Sun.

Beta Pictoris has one too

Then, in 1984, scientists at an observatory in Chile actually photographed a similar disk around the star Beta Pictoris (see page 248). How many other Vegalike systems are out there? IRAS studied the 9,000 brightest stars in the sky and found only 50 other candidates, so these large dust rings are not very common. Within the next two decades astronomers using new techniques will be able to search all the nearest stars to find out how many possible planetary systems there really are.

Hubble troubleshooters work in a 1.3-million-gallon tank of water to simulate some of the conditions they will find while working in space. Astronauts train for the NASA mission to repair the flawed Hubble Space Telescope. Here they are rehearsing the exchange of the telescope's old solar panels for new ones. On the repair mission, in December 1993, shuttle-borne astronauts made five space walks, replacing Hubble's faulty parts and giving it one new camera and new mirrors for another. Now the telescope sends back sharp pictures from deep space. Astronomers hope to use it to search for other planets and to study black holes.

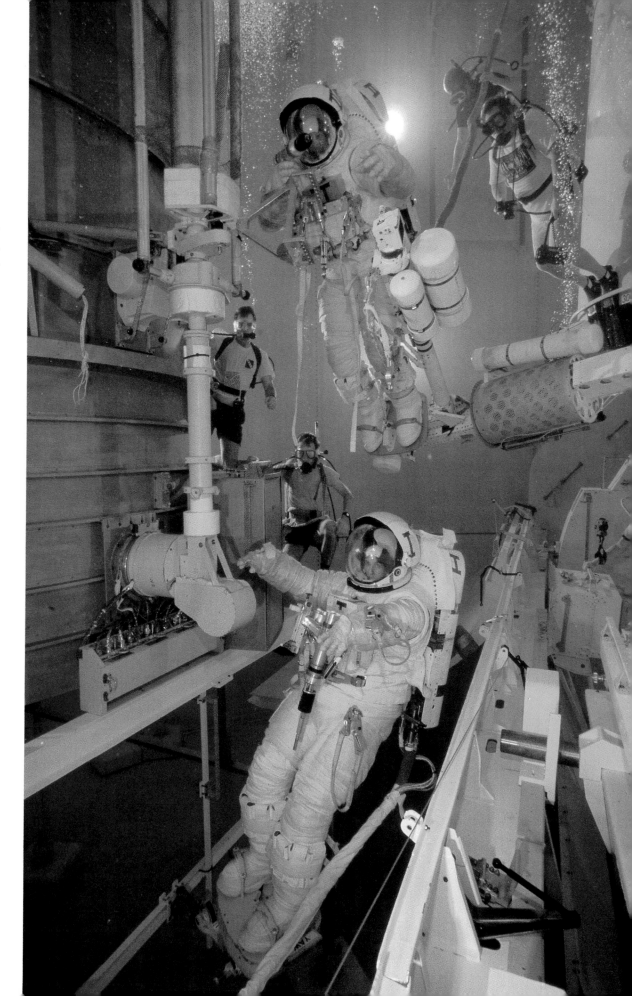

How did life on Earth begin? Chemist David Usher revolves a heat lamp around an organic brew to simulate early Earth's conditions. Such experiments, pioneered by Stanley Miller, try to duplicate the chemical events that transformed life's raw ingredients into living cells.

Another planet with humans?

Can we expect other life, if it exists out there in the Universe, to be like ours? We might expect to find certain general likenesses based on the biological principles we know—the kinds of molecules that make up an organism and the way the organism's chemistry works. Although we might expect to find organisms that walk on legs, swim, crawl, or fly, we should not expect to find pigeons, dogs, or humans on any other planet in the Universe.

Like Earth, other planets with life would change continually through the ages. And no two are likely to change in the same way. This means that evolution is not likely to follow identical pathways on any two planets, or ever to repeat the same pathway on any one planet. No, it is almost certain that there cannot be even one carbon copy of Earth life anywhere else in the Universe. There may be intelligence, love, even wisdom, but nowhere should we expect it to wear a human shape.

"Hello, out there"

If there are other intelligent beings out there in the dark, will we ever be able to communicate with them? And if so, how?

Radio telescopes can serve as our interstellar telephones. At radio frequencies, cosmic "noise pollution" from all the sources of radiation in the Universe is lowest. Radio signals—like all electromagnetic radiation—travel at the speed of light. Since our cosmic "hello" cannot travel faster than light, we would have to beam a message at a nearby star for there to be any chance of receiving a reply in our lifetime.

If intelligent beings are out there, maybe they will learn about us someday. In 1974 the giant radio telescope at Arecibo, Puerto Rico (right), sent a coded message into space. Shown here in a decoded image, the message begins at the top with binary numbers 1 to 10 and the atomic numbers of major elements. Green, white, and blue describe DNA, the molecule that determines heredity. Marks on either side of the human figure tell its height and Earth's population. The Solar System is yellow. Outlined at the bottom is the Arecibo telescope with its size.

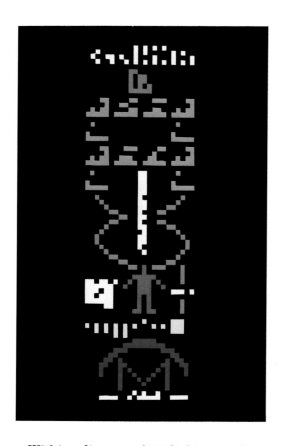

Within a distance of 100 light-years from us there are some 10,000 stars. But at that distance it would take 200 years to exchange a message. Within 20 light-years from us are 22 stars like the Sun; some may have Earthlike planets. The nearest stars to us are those of the Alpha Centauri system. It would take our message four years and four months to reach Alpha Centauri, then another four years and four months for an Alpha Centaurian to answer.

A much more efficient strategy for communicating with outer space is to *listen* for a signal. This could yield an immediate result. For three months in 1960 a radio tele-

Voyager 1 swings by Saturn in 1980, carrying a message encased on its side. If spacefarers encounter Voyager among the stars, they can see electronic photographs and hear Earth sounds: Rain, surf, a train whistle, laughter, a kiss, barking dogs, music, greetings in 55 languages.

scope at the National Radio Astronomy Observatory at Green Bank, West Virginia, was tuned in to the nearby stars Tau Ceti (11.9 light-years away) and Epsilon Eridani (10.7 light-years). The astronomer in charge, Frank Drake, named his project Ozma, "for the princess of the imaginary land of Oz, a place very far away, difficult to reach, and populated by exotic beings." During his short listening period Drake did not hear a single intelligent whisper.

The cosmic haystack

Since 1960 other listening posts have been set up, but they have barely begun to explore all the possible combinations of signal frequency and modulation, time of search, and strength and direction of telescope. Much more could be done, but even so, expecting to pick up an interstellar greeting is like expecting to find that needle in the haystack.

Although our television and radio signals continually leak into space, advertising our presence, we have sent only one message intentionally. That was in 1974 when the Arecibo message (page 251) was beamed to the Great Cluster in Hercules, 21,000 light-years away. If alien radio astronomers answer, their reply will not reach Earth for 42,000 years.

Meanwhile we keep listening to likely stars in hopes of receiving a greeting — perhaps from a planet thousands of light-years away, whose civilization no longer exists. No matter. If we ever do receive such a message it will be the most exciting event in history, for it will tell us that we are not alone in the Universe.

Shuttles & Starships

Taking off as a rocket, it orbits Earth as a spaceship, and lands back on Earth as a heavy glider. It's the 37-meter-long space shuttle, a cosmic pickup truck designed to carry almost 30 tons of communications or weather satellites, a 9-ton space telescope, and other spacecraft, and place them in orbit around Earth.

The astronaut crew member of the orbiting shuttle shown here is maneuvering into position to perform maintenance chores on a telescope satellite. A backpack equipped with small thruster jets enables the astronaut to move in any direction with precision.

Far in the future, shuttle-like spacecraft will have many tasks. They probably will ferry assembled parts of Earth-orbiting space cities housing a thousand or more people, and components of settlements on the Moon, Mars, and moons of Jupiter and Saturn. Later will come the design and construction of spacecraft able to cross the still greater distances to remote Neptune and ice-encrusted Pluto and return. These planets, too, are bound to be explored during the coming century.

Is this idle dreaming? Perhaps it would have been, a hundred years ago, but not today. We now have the technology for such

A thunderous blast-off from Cape Canaveral, Florida, sends a NASA shuttle into orbit, piggyback on its fuel tank. Part spaceship, part glider, the reusable cargo ship can carry a load as heavy as a full truck into space. Space travel began a new era with the first shuttle flight in 1981.

ventures. We have had a taste of space — and along with its rewards we have suffered its perils. Brave men and women from several nations have dared to explore and some have given their lives. But the response to setbacks is always to go forward. The will to explore cannot be denied.

On January 28, 1986, the space shuttle *Challenger* stood on the launch pad as its crew of seven men and women braced for liftoff. Behind them in time stretched 25 years of spectacular achievements in human space flight. Astronauts had walked in space and on the Moon and lived in orbiting capsules for months at a time. And four shuttles had flown a nearly flawless series of 24 missions that began with the three-day flight of *Columbia* on April 12, 1981.

Ignition. Liftoff. Black smoke puffed from the side of *Challenger*'s right booster. A minute later a jet of flame burst from the same area — and then the world stared in disbelief at the twisted cloud that hung in the Florida sky. A mighty explosion had claimed shuttle and crew and brought the manned space program to a halt until scientists could make sure that whatever happened would never happen again.

Made in space

Challenger had been the workhorse of the fleet. On one of its nine earlier missions, its crew produced the first made-in-space items to sell — nearly a billion tiny plastic beads of the type that scientists need in microscopes to make detailed studies of blood cells. But why make beads in space?

The answer: weightlessness. Here on the ground, gravity holds things down. If it

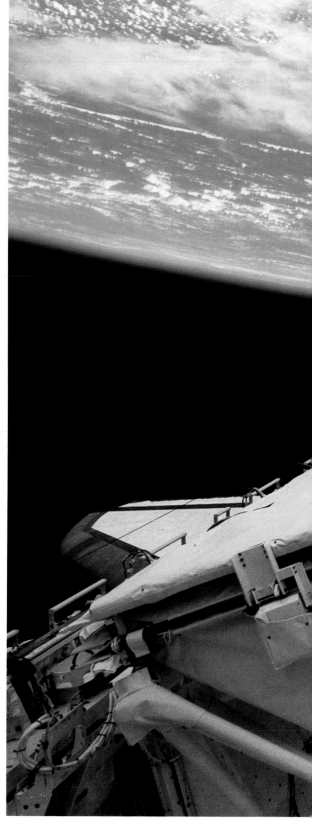

Which way is up? For shuttle astronauts
in orbit around Earth the absence of
gravity is all part of the daily routine.
David Leestma (opposite in red stripes)
and Kathryn Sullivan seem to stand on
their heads in the cargo bay of NASA's
Challenger as they change weightless fuel
from tank to tank. Weightlessness also
makes eating an adventure: Richard
Mullane, sitting on the thin air of shuttle
Discovery, catches a shrimp as other crew
members of the September 1984 flight
juggle dinner. Aboard another space
shuttle in 1985, Wubbo Ockels of the
Netherlands adjusts a helmet in the
European Space Agency's Spacelab to test
the effect of weightlessness on balance.

259

didn't, the outward pull of centrifugal force would fling us off our spinning globe like pebbles off a whirling bicycle tire. But a capsule in orbit speeds around much faster than the Earth spins, so its centrifugal force nearly cancels out gravity. The capsule and everything in it float in zero g, or zero gravity.

On the ground a bead of melted plastic is distorted by its own weight as it hardens, just as a balloon full of water is pulled into a teardrop shape. But in zero g, beads form nearly perfect spheres of the same size.

Working in space

There's more to do in a shuttle than make beads. Even before *Columbia's* first flight, governments, universities, and private companies had booked the first 40 flights to launch satellites, do experiments, and test ways of making things in space. School students designed some of the experiments; one idea sent a spider into space to see whether she would spin a normal web in zero g. She did.

Some flights carried Spacelab, a fully furnished laboratory built by 11 European countries, with instruments for studying such things as the Sun, stars, cosmic rays, and Earth's ionosphere. Scientists on the projects, as well as career astronauts, work in Spacelab. They are called payload specialists. In one Spacelab project, payload specialists studied rats and monkeys for motion sickness, which half the astronauts have felt. A monkey got sick too—but all 24 rats apparently loved the ride.

When a car stalls, you fix it. But when a satellite that cost $77,000,000 breaks

down 480 kilometers up—and a shuttle pulls alongside and fixes it—that's road service at a new high. On April 8, 1984, *Challenger* parked beside Solar Max, a car-size satellite that studied the Sun until its aiming controls failed. An astronaut in an MMU—a Manned Maneuvering Unit like the one on page 254—spacewalked over to Solar Max but couldn't tow it to the shuttle. So the shuttle came to Solar Max, inching closer until its Remote Manipulator Arm could grab the satellite and pull it into the cargo bay. For seven hours, the crew tinkered with the aiming controls and made other repairs. Then *Challenger* put Solar Max back in orbit, good as new.

Later, the shuttle *Discovery* made a service call on the ailing Syncom 3 communications satellite. This time an astronaut rode out on the end of the jointed arm to grab the satellite. Fixing space hardware has become a big part of the shuttle program's job—a function the astronauts dubbed the "Ace Satellite Repair Co."

Giant steps toward space

Three giant steps in space exploration began in a shuttle's cargo bay. The Hubble Space Telescope emerged into orbit, free of Earth's hazy air, to peer seven times deeper into space than ground telescopes can. Bigger than a bus, the instrument marks one of the greatest advances in astronomy since the work of Galileo. It may even show us what the Universe was like when it formed 15 billion years ago.

The work of another Galileo—this one an unmanned probe—began with a ride in a shuttle and a journey to Jupiter to study

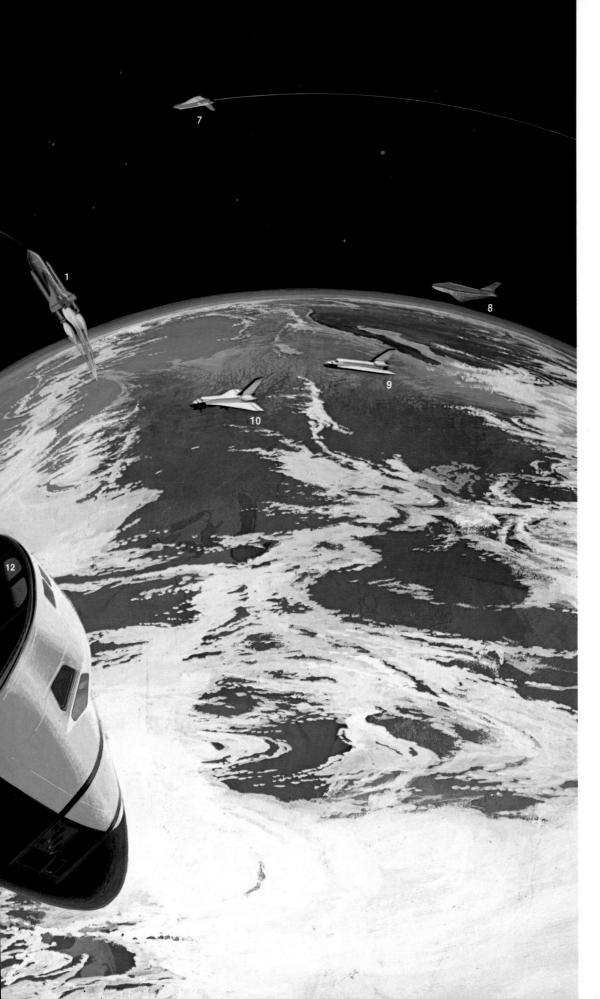

its atmosphere and moons. And a European Space Agency probe called Ulysses streaked from a shuttle's Earth orbit into orbit around the Sun. It passed close to Jupiter, where the giant planet's gravity flung it into an orbit that puts it over the Sun's poles, a maneuver dubbed a "gravitational slingshot." From its unique vantage point above and below the Solar System's plane, Ulysses will tell us about the Sun's magnetic field, the solar wind, and the matter and energy to be found in "empty" space. Speed champion of spacecraft, Ulysses zips along at about 92,000 kilometers an hour. Faster than a speeding bullet? About 20 times faster!

What's next? Space stations, orbiting

A space shuttle should make as many as 100 missions, in about this sequence: (1) Seven seconds after launch, shuttle begins roll to tail-down attitude. (2) In 2 minutes, boosters separate about 50 km up. (3) Boosters parachute to sea for recovery. (4) Main engines continue to about 110 km up. (5) After 8 minutes, main engines stop; empty tank separates; small engines thrust craft into orbit. (6) Tank falls into Indian Ocean. (7) Orbiter coasts in equatorial orbit about 275 km up. (8) Mission complete, parts of orbiter glow with heat of entry into atmosphere. (9) Orbiter glides toward landing site. (10) Wheels extend just before touchdown. Crew in orbit (11) works in pressurized spacelab. Cabin and flight deck (12) connect with lab by tunnel. On another mission, orbiter's manipulator arm (13) lifts satellite from hold, puts it in orbit.

263

Shedding its red-hot heat shield, a probe plummets into Jupiter's clouds. It will sample the atmosphere until crushed by the pressure. A second craft in orbit relays data to Earth. Named Galileo, the probe—like the space telescope pictured here—was launched from a space shuttle.

factories, and laboratories fill NASA's long-range plans. And when today's shuttles retire, a new generation of reusable space trucks will take to the skies, carrying parts for new space stations and astronauts to assemble them.

American astronauts have for many years practiced construction techniques to use on a space station. In 1985 crew members of the shuttle *Atlantis* easily assembled 93 aluminum tubes and 33 joints into a 14-meter-long girder like those that might be used to build a space outpost. Russian cosmonauts did welding in space to repair parts of their space station, Mir. Now the United States, along with Europe, Japan, and Canada, hope to work with Russia to form an international space station and to learn how to live and work in space for long periods. Perhaps such cooperation will lead to joint missions back to the Moon and on to Mars.

A city in the sky

In 1975 some 30 scientists met for ten weeks to design a city in space. Using millions of tons of raw materials from mines on the Moon and asteroids, they proposed to build a gigantic circular tube called a torus, a wheel nearly two kilometers across (pages 268-269). Inside, 10,000 people would lead normal lives, working, playing, raising families and flower gardens, going to schools, theaters, and sports arenas. The colonists would breathe an Earthlike atmosphere and, because the wheel rotates once a minute, they would feel a normal Earth gravity produced by its spin. Their city in the sky would be self-sufficient.

Space colonists could build solar power

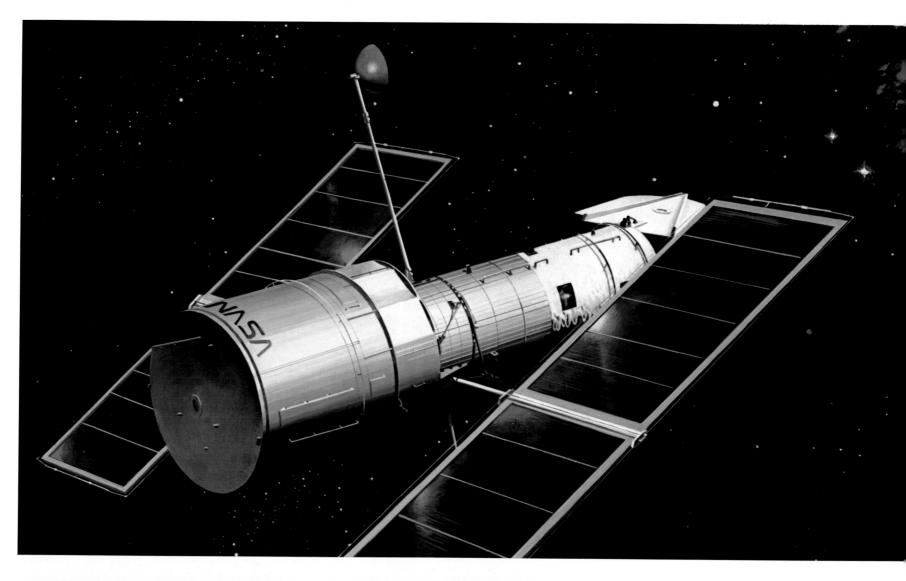

In orbit 500 km up, the Hubble Space Telescope opens its eye to wonders unseen by Earth telescopes. With no atmosphere to mar its view, it can study 350 times the volume of space ever seen before, peering to the edge of the Universe. With shuttle crews servicing it, the space telescope should last at least 20 years.

In 1986 spacecraft from several nations flew by Halley's Comet to examine the chemical and physical makeup of its coma, nucleus, and tail. To learn more in the future, a comet-chaser will greet its target with open arms, each covered with solar cells that power its electrical system.

On the Moon's far side, a research base seeks data above and below. Free of Earth's radio noise, dish antennas listen for signals from space. An observatory scans the airless sky. Geologists drill rock samples, then ride a rover to homes and labs buried in protective coats of dirt.

Prospectors of the 21st century hitch up to an asteroid and scout its mineral wealth. They may mine it on the spot, or tow it to a space factory better equipped to extract its deposits of metals, carbon, possibly water—necessary raw materials for a zero-gravity manufacturing plant.

satellites big enough to supply all the energy Earth needs. Space factories at the wheel's hub or floating nearby could make a host of articles to near-perfection in zero-g. Enormous optical telescopes and radio telescopes many kilometers across could be assembled. Spacecraft cities could be built to carry people far from Earth. Some of these sky-islands might explore the ultimate frontier—intergalactic space.

Breaking the time barrier

The laws of physics tell us that there is a speed limit we cannot exceed or even reach—the speed of light. Since light from

even some nearby stars takes many years to reach us, a starship traveling at the snail's pace of today's fastest spacecraft would take millions of years to journey to one of those nearby stars and return.

Right now we know of no way to break this barrier. But perhaps there is a way of cheating time. Evidence suggests that, as space travelers approach the speed of light, time will slow down for them. This is *time dilation,* part of Einstein's special theory of relativity. Suppose a starship crew voyaged to the star Mira, about 100 light-years away. If they traveled at 99 percent the speed of light they would make the

267

Cities in space may look like big bicycle wheels, each with a free-floating mirror to gather in sunlight (left). High-rise spokes contain shops, labs, and offices. Transport tubes whisk the city's 10,000 residents between homes in the outer tube—or torus—and work and play areas at the hub. A long tube links the wheel with a spacecraft dock, power plant, and factory unit. Like tread on a tire, Moon rock shields the torus from meteoroids and radiation. Inside, farm workers adjust water and nutrients (below) for crops ripening in soil-less racks.

round trip in only 28 years their time. But back home they would find their friends were 200 years older. Because of time dilation, starship crews may explore the far reaches of the Milky Way, and even distant galaxies, and return in their lifetime.

Let's go back to one of our space settlements that has existed near Earth for several generations. We find that the inhabitants have voted to embark on stellar travel. They decide to speed up gradually at a comfortable acceleration rate of one g, or one unit of gravity, which is the amount you feel when you stand still on Earth's surface. When you take off in a jetliner you are pushed back in your seat with a force of less than one g. When the Apollo astronauts re-entered Earth's atmosphere they felt about six g's. For a short time that isn't harmful, but it is very uncomfortable—your face sags, your skin pulls, your arms and legs feel heavy as lead. Accelerating the starship at one g will cause no problem for our space travelers.

Setting out for the stars

They decide on a round trip to the star Epsilon Eridani, a yellowish pinpoint in the sky. As the spaceship heads toward Epsilon Eridani, it will speed into the light waves from the star like a boat speeding through waves on a lake. The frequency of the light waves (the rate at which they pass the starship) will increase as the craft speeds up. The crew will see the light from Epsilon Eridani blue-shifted, or shifted toward the violet end of the spectrum. This is the Doppler effect (pages 28-29 and 245). The light from the yellowish star will shift

Boxy robot craft called wardens tend the fuel globes of an unmanned Daedalus starship. Two Enzmann starships wait nearby, each with its huge ball of frozen deuterium linked by manned modules to nuclear-pulse engines. Two designs, one goal: to reach a nearby star.

through higher frequencies into the green band, then into blue, then violet.

At still faster speeds, Epsilon Eridani will be seen to shine with a pure violet light. Then it will fade as its light shifts off the visible spectrum into the ultraviolet. The star has become almost invisible, although it is still there.

Just the opposite happens when the travelers look homeward toward the Sun. Its light waves are not being crowded up. They are being stretched out to lower frequencies and are red-shifted. Gradually the familiar yellowish Sun will appear orange and then deep red. Then it will fade from view as its light shifts off the visible spectrum into the infrared. The Sun, too, has become almost invisible.

Epsilon Eridani is 11 light-years from Earth, so the trip out and back will take 10 years, starship time. But on Earth 22 years will pass. Longer journeys at such speeds would distort time even more. To travel to the center of our galaxy, 23,000 light-years away, and back could take 35 years starship time but 46,000 years Earth time. The Andromeda Galaxy is two million light-years away. To travel there and back could take 55 years by starship while on Earth four million years would go by.

How would it feel to come "home" to an Earth that had aged four million years in your 55-year absence? Earthlings might have evolved beyond recognition and could regard you as a dangerous alien life form. But scientists among them would be fascinated to talk to scientists on your starship, to compare species and geological forms from four million years earlier to things as

they appeared "today." Such a starship would be a magnificent time machine.

To "elsewhere and elsewhen"

If starships do journey to distant corners of the Universe, most likely the travelers will say goodbye to Earth forever. They will spend their lives comfortably within their space capsule. They will visit stars likely to have habitable planets or civilizations. The astronomer Carl Sagan wonders if tomorrow's space explorers will be whisked from one part of the Universe to another through black holes. Sagan suggests black holes could be entrance ways to "elsewhere and elsewhen." He speaks of a "federation of societies in the Galaxy that have established a black hole rapid-transit system. A vehicle is rapidly routed through an interlaced network of black holes to the black hole nearest its destination."

Can we reach the stars?

Some scientists think we will never reach the stars; time will forever keep us prisoners of our Solar System. But people once scoffed at the idea of huge machines carrying passengers through the air and of trips to the Moon. Yesterday's science fiction often becomes today's science fact.

We have many reasons to venture into space. One is to harness energy. Another is to find out about other planetary systems and stars so we can better understand our own. And finally there is the driving curiosity to know whether we are alone in the vastness of the Universe. Do intelligent beings and advanced civilizations abound out there? One day, perhaps, we will know.

Space Age Highlights

1903 Konstantin Tsiolkovsky, astronautics pioneer in Russia, publishes an article on spaceflight, the first mathematical proof that travel by rocket is possible.

1905 Albert Einstein (above) proposes his special theory of relativity, a description of the structure of space and time that revolutionizes science.

1916 Karl Schwarzschild, a German physicist, introduces the modern black hole concept. French astronomer Pierre Laplace first suggested the idea in 1796.

1917 U. S. astronomer Harlow Shapley determines the true size of our galaxy. This discovery put the Sun not at the center, but 30,000 light-years out toward the edge.

1924 Astronomer Edwin P. Hubble proves galaxies exist outside the Milky Way. Hubble later studies galaxies with the powerful Schmidt Telescope (above) at Mount Palomar Observatory in California.

1926 In Auburn, Massachusetts, physicist Robert H. Goddard launches the first successful liquid-fuel rocket (above).

1927 Belgian cosmologist Georges Lemaître formulates the Big Bang theory. This proposes that all matter and space— the entire Universe—was born out of the colossal explosion of a "primeval atom."

1929 Edwin P. Hubble finds that the more distant a galaxy, the faster it recedes. Hubble's Law demonstrates that the Universe is expanding.

1930 At Lowell Observatory, in Flagstaff, Arizona, Clyde Tombaugh discovers the ninth planet, Pluto, by following the calculations of Percival Lowell.

1931 American pioneer of radio astronomy, Karl Jansky (above), discovers radio waves coming from the Milky Way Galaxy.

1937 Radio astronomer Grote Reber follows up Karl Jansky's work and builds the first "dish" radio telescope.

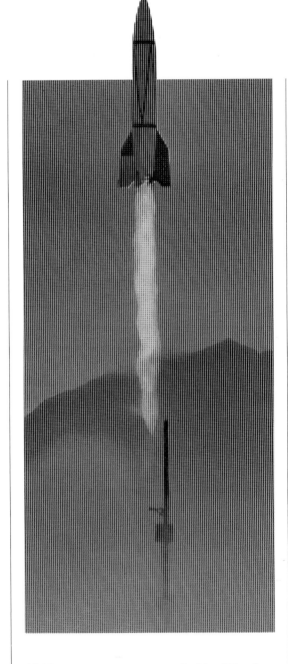

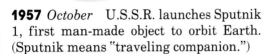

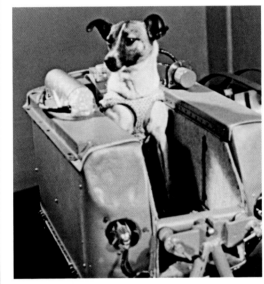

1957 *October* U.S.S.R. launches Sputnik 1, first man-made object to orbit Earth. (Sputnik means "traveling companion.")

November Sputnik 2 transports the first living creature into space. Confined in a small compartment, the dog Laika lives in orbit for seven days, proving that life can survive on "the final frontier."

1958 *January* Explorer 1, first U. S. satellite in orbit, lifts off at Cape Canaveral. Geiger counters on board confirm the existence of Van Allen radiation belts.

1959 Soviet probes make space history. Luna 1 is the first spacecraft to orbit the Sun. Luna 2, first to reach another celestial body, hits the Moon. Luna 3, aloft for 177 days, is the first to return pictures of the lunar far side.

1960 *April* U. S. launches Tiros 1, the first successful weather satellite.

1961 *April* Soviet Vostok 1 carries the first man in space, cosmonaut Yuri A. Gagarin (above), once around Earth.

May Alan B. Shepard, Jr. (above), the first U. S. astronaut in space, makes a 15-minute suborbital flight in Mercury capsule Freedom 7, to an altitude of 187.5 km.

1942 German scientists, led by Wernher von Braun, build the first successful V-2 rocket. After World War II, a captured V-2 (above) boosts the U. S. WAC Corporal to a record height of 402 km at the White Sands Proving Ground, New Mexico.

1962 *February*   John H. Glenn, Jr., the first American in orbit, circles Earth three times in Friendship 7, reaching 261 km altitude. Celebrations greet his return. At Cape Canaveral he rides in a parade with President John F. Kennedy (above).

July   U. S. satellite Telstar 1 beams the first live transatlantic telecast (above), from the U. S. to Europe.

December   The first successful planetary spacecraft, U. S. Mariner 2, flies past Venus at a distance of 33,635 km.

1963 *June*   Soviet cosmonaut Valentina Tereshkova sorts gear (above) after her flight as the first woman in space. In Vostok 6, she orbits Earth 48 times.

Maarten Schmidt at Mount Palomar interprets the unusual behavior of "radio star" 3C 273 — the first known quasar (above).

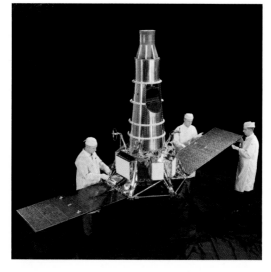

1964 *July*   U. S. lunar spacecraft Ranger 7 relays the first close-range photographs of the Moon. The craft undergoes final inspection in California (above), before being shipped to Cape Canaveral for launching.

November   Mariner 4 is launched by the U. S. and returns the first detailed data about Mars (above). From 9,846 km away, Mariner reveals Mars' surface craters.

February The U.S.S.R. Luna 9 soft-lands on the Moon and relays the first pictures directly from the lunar surface.

March The U.S.S.R. launches Luna 10, the first spacecraft to orbit the Moon.

June Surveyor 1, the first U.S. spacecraft to soft-land on the Moon, takes a picture of its own shadow (above).

1965 *March* Cosmonaut Alexei Leonov makes the first space walk, from the U.S.S.R. Voskhod 2. Duration of EVA (extravehicular activity) is 10 minutes.

June James McDivitt pilots Gemini 4 as Edward White II (above) takes the first U.S. space walk. EVA is 21 minutes.

December Walter Schirra, Jr., and Thomas Stafford, U.S. astronauts, make the first rendezvous in space, maneuvering Gemini 6 alongside Gemini 7 (right).

December Frank Borman and James Lovell, Jr., complete 206 Earth orbits in Gemini 7. The 14-day flight proves that travel to the Moon and back is possible.

1967 In Cambridge, England, Antony Hewish (above) thinks at first of messages from outer space when his student, Jocelyn Bell, receives radio pulses from pulsars.

275

1968 *September* Soviet Zond 5 is the first spacecraft to orbit the Moon and return to Earth. Its cargo includes plant and animal life to test radiation danger in space.

December U. S. launches Apollo 8, the first manned spacecraft to orbit the Moon.

1969 *July 20* Neil Armstrong and Edwin Aldrin, Jr., leave the first footprints on the Moon (above), as Michael Collins orbits in Apollo 11's command module. Eight days after blast-off (right), they return to Earth with samples of Moon rock (below).

1970 *December* U.S.S.R. Venera 7 is the first probe to soft-land on Venus. It transmits atmospheric data from the surface, confirming high temperature and pressure.

1971 *July* U. S. Apollo 15 astronauts David Scott and James Irwin drive the first Moon rover. A year later, Apollo 17's Harrison Schmitt, the first geologist-astronaut, mans a similar rover (above), which travels 35 kilometers during the mission.

November U. S. Mariner 9 is the first spacecraft to orbit another planet. For almost a year, it takes more than 7,000 pictures to help map the Martian surface.

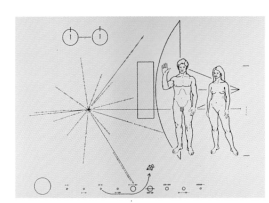

1972 *March* U.S. launches Pioneer 10 toward Jupiter and interstellar space. Designed to acquaint alien life with Earth civilization, the Pioneer plaque (above) shows human figures, the spacecraft's trajectory on a Solar System diagram, and a pulsar map to help fix the time between takeoff and discovery.

December Using information from the X-ray satellite Uhuru, scientists designate Cygnus X-1 the first probable black hole.

1973 *May* U.S. launches the Skylab space station. In Earth orbit 435 km up, Skylab crews work and live in zero gravity (center). Projects range from solar studies to astro-spiders Anita and Arabella, who show how to spin webs in space.

November U.S. Mariner 10 undertakes a double-planet mission. The spacecraft flies past Venus and on to Mercury, giving scientists their first good look at Mercury's cratered surface.

1974 *December* Pioneer 11 travels by Jupiter en route to Saturn, where five years later it discovers and photographs new moons and additional rings.

1975 *October* Venera 9 and Venera 10 transmit to Earth the first pictures of the Venusian surface. Findings classify Venus as a still-evolving planet.

1976 *July* Pictures of the rocky, red surface of Mars are returned by U.S. Viking 1. Soil samplers (right) search for life in the Martian soil but find none.

1977 *August-September* Voyagers 1 and 2 leave Earth for a 1979 date with the planet Jupiter and 1980 and 1981 meetings with Saturn. In January 1986 Voyager 2 flew by Uranus. When Voyager 2 reached Neptune in August 1989, it became the first spacecraft to fly by four planets.

1978 *December* Two Pioneer spacecraft reach Venus. One drops four probes into the atmosphere; the other, an orbiter, maps the surface with radar.

1981 *April* The first flight of space shuttle Columbia inaugurates a new era in space travel.

1983 *January-November* The Infrared Astronomical Satellite (IRAS) finds new comets, asteroids, galaxies, and, around the star Vega, a dust ring that may be a new planetary system forming.

1986 *March* Spacecraft from the Soviet Union, Japan, and Western Europe fly by Halley's Comet on its 30th recorded appearance.

1986 *March* Astronomers discover an invisible gravity source beyond the Milky Way that is so powerful it bends and splits a distant quasar's light.

1988 *December* Soviet cosmonauts aboard the space station Mir set a 366-day record in space.

1992 *October* NASA launches SETI—the Search for Extraterrestrial Intelligence. Today SETI and other such endeavors operate on private donations.

Glossary

Most scientific terms used in *Our Universe* are explained in the text and appear in the Index. This Glossary lists some of the most important words with short definitions for quick reference.

Absolute magnitude: a measure of the true brightness of a star as if all stars were the same distance (32.6 light-years) from the observer.

Antimatter: material made of atomic particles having certain properties opposite to those of ordinary matter. If combined, matter and antimatter would explode.

Apparent magnitude: the brightness of a star as seen from Earth.

Apparent motion: the movement of a celestial body against the background of distant stars.

Asteroid: a rocky object, smaller than a planet, that orbits the Sun.

Astrology: a nonscientific system that attempts to explain or predict human actions and events by the position of celestial objects.

Astronomy: the branch of science that studies the Universe beyond Earth's atmosphere.

Atom: the smallest possible unit of a chemical element; an atom consists of a nucleus and one or more orbiting electrons.

Axis: an imaginary line around which a body, such as a planet, rotates.

Binary star: two stars orbiting a common center of gravity.

Black dwarf: a white dwarf which has stopped radiating energy.

Black hole: in theory, a collapsed object—perhaps a massive star—whose gravitational field is so strong that under most circumstances no light or matter can escape.

Comet: a small body made of ice and dust which orbits the Sun.

Compound: a substance formed by the chemical combination of two or more elements. Hydrogen and oxygen combine to form water.

Constellation: a pattern of stars; one of 88 areas dividing the sky.

Convection: the vertical movement of energy or mass by circulating currents.

Core: the innermost part of a moon, planet, or star.

Cosmic rays: nuclear and subatomic particles moving through space at high speeds; radiated from the Sun and other stars.

Cosmology: the study of the origin, evolution, and overall structure of the Universe; creation myths and the Big Bang are theories of cosmology.

Crust: the thin, outermost rocky layer of a moon or terrestrial planet.

Density: a measure of how tightly mass is packed into a given space.

Doppler effect: change in frequency of sound or light waves caused by relative motion of the source and the observer.

Eclipse: total or partial blocking of light from a celestial body caused by its passing into the shadow of another body; also, the hiding of one celestial body by another.

Ecliptic: the Sun's apparent circular path through the heavens; the plane of Earth's orbit.

Electromagnetic spectrum: the whole range of radiation; it extends from high-energy gamma rays to low-energy radio waves.

Electron: a negatively charged, low-mass particle which orbits the nucleus of an atom or exists free in space and in stars.

Element: the simplest form of chemical, made of identical atoms. Oxygen and gold are among the more than 100 known elements.

Energy: the ability to do work, such as produce motion, heat, chemical change.

Frequency: rate at which light or sound waves pass a given point, measured in cycles per second, or hertz. Shortwaves pass the point at high frequencies, longer ones at low frequencies.

Galaxy: billions of stars held together by gravitational attraction.

Gas giants: the planets Jupiter, Saturn, Uranus, and Neptune.

Giant: a large star that is highly luminous.

Globular clusters: dense groups of thousands to millions of stars.

Gravitation: the force by which two masses attract each other.

Gravity: gravitational force at the surface of a planet or other body that pulls mass toward its center.

Interstellar dust: small, solid grains of matter, thinly distributed between the stars; sometimes concentrated in nebulae.

Interstellar gas: matter in a gaseous state thinly distributed between the stars—mostly hydrogen with some helium and other elements; sometimes concentrated in nebulae.

Ion: an atom that has lost or gained one or more electrons, becoming positively or negatively charged.

Limb: the edge of a planet or other celestial body as seen from afar.

Luminosity: a measurement of the total amount of energy given off from a star.

Mantle: the layer of rock between the core of a moon or terrestrial planet and its surface crust.

Mass: the amount of matter contained in a body.

Matter: the substance of the Universe; made up of atomic particles, atoms, and molecules. Matter exists in four familiar states: solid, liquid, gas, and plasma.

Meteor: the streak of light caused by a meteoroid that passes through Earth's atmosphere; the burning meteoroid, called a "shooting star."

Meteorite: a meteoroid that survives passage through Earth's atmosphere and arrives on Earth's surface without completely burning up; also, a chunk of rock after it lands on a planet or moon.

Meteoroid: a solid body smaller than an asteroid that orbits the Sun. Both meteors and meteorites were meteoroids.

Molecule: two or more atoms chemically combined. A water molecule consists of two atoms of hydrogen and one atom of oxygen.

Moon: a natural satellite orbiting a planet.

NASA: National Aeronautics and Space Administration.

Nebula (pl. nebulae): a mass of interstellar gas and dust, appearing as a glowing or dark patch in the sky.

Neutrino: an atomic particle given off during nuclear fusion; it apparently has little or no mass and moves at the speed of light.

Neutron: an atomic particle having high mass but no electrical charge. Neutrons are present in the nuclei of all atoms except hydrogen.

Neutron star: the core of a star left after a supernova explosion; made of very densely packed neutrons.

Nova (pl. novae): a star which suddenly explodes, temporarily increasing its brightness.

Nuclear fusion: a process by which matter changes to energy. The nuclei of lighter atoms fuse, or join, to form heavier nuclei, releasing energy.

Nucleus (pl. nuclei): of an atom, the central portion which has a positive charge and contains most of the atom's mass; of a comet, the chunk of solid matter at the center of a comet's head; of a spiral galaxy, the central portion, very dense with stars.

Occultation: the hiding of one celestial body by another.

Particle: any very small piece of matter such as a molecule or atom; also pieces even smaller such as electrons, protons, and neutrons; also larger ones, as in interstellar dust. Photons are particles of light.

Phases: of a planet or moon, the varying shape of the lighted portion, such as full, half, crescent, etc.

Photons: particles that make up electromagnetic radiation. Photons carry varying amounts of energy and travel at the speed of light.

Planet: a rotating body of substantial size held in orbit by the gravitational attraction of a star. A planet is not self-luminous; it reflects starlight. Its own gravity pulls a planet into its most stable shape, a slightly flattened sphere.

Plasma: a gas consisting of electrons and ions; called "the fourth state of matter" because the temperature is too high for whole atoms to exist.

Precession: a slow change in the direction of the tilt of Earth's axis, which results in an apparent change in the position of the stars.

Pressure: a measure of the force exerted on a surface.

Prism: a wedge-shaped piece of glass used to split light into a spectrum.

Proton: an atomic particle with high mass and a positive charge, present in the nuclei of all atoms. The nucleus of a hydrogen atom is a single proton.

Pulsar: a neutron star that rotates rapidly and emits a beam of radiation. Earth telescopes pick this up as a regular pulse.

Quasar: a mysterious, "quasi-stellar," or starlike, object in a galaxy's core—very small, very bright, very distant. Most quasars are strong sources of radio energy.

Radar: radio signals transmitted to and bounced back from an object. RADAR is an acronym for RAdio Detecting And Ranging.

Radiation: energy transmitted through space as waves or particles.

Red dwarf: a small, relatively cool star with low luminosity.

Relativity: Theories of physics proposed by Albert Einstein. They say, among other things, that space and time cannot be considered separate ideas. The perception of space-time is different for a person standing still on Earth than it is for someone moving very fast away from or toward it. What we see is *relative* to (depends on) our acceleration as we move.

Retrograde motion: "backward"; opposite of direct motion; the apparent east to west motion of a planet as seen from Earth; clockwise revolution or rotation of an object, as seen from above the north pole.

Satellite: a moon or man-made body in orbit around a planet.

Silicates: a large group of minerals containing silicon and oxygen, usually combined with one or more metals. Most common rocks are silicates.

Solar wind: a stream of charged particles from the Sun.

Solstice: the moment when the Sun is at its northernmost or southernmost point in the sky; the first day of summer or winter.

Star: a hot, glowing sphere of gas, usually one that emits energy from nuclear reactions in its core.

Supergiant: a large massive star of low density and very high luminosity.

Supernova: a stellar explosion which increases the luminosity of the star to many thousands of times brighter than a nova.

Terminator: The shadow-line boundary between night and day on a planet or moon.

Terrestrial planets: Mercury, Venus, Earth, and Mars.

Tides: periodic changes in the shape of a planet, moon, or star caused by the gravitational attraction of a body near it. For example, the Moon tugs on Earth's oceans, causing high and low tides; Jupiter's gravitational attraction on its moon Io causes ground tides; and two stars very close together pull each other's atmospheres into distorted shapes.

Transit: the crossing of a small celestial body in front of a larger one. From Earth we see Venus and Mercury in transit across the disk of the Sun.

Vacuum: in theory, space that contains no matter.

Variable star: a star whose brightness changes over time.

Wavelength: the distance between two successive crests of a wave.

White dwarf: a type of star that has collapsed after exhausting its nuclear fuel. Leftover heat causes it to shine faintly.

Measuring the Universe

Our Universe gives measurements of length, weight, and area in metric units because scientists use the metric system in their work. In *Our Universe* you will find temperatures measured in degrees Celsius (°C), also called centigrade, and in kelvins (sometimes abbreviated K). Kelvins are used by astronomers to describe very high temperatures, such as those in the Sun and other stars. The equivalents given below will help to explain terms of measurement used in this book.

Length

1 millimeter (mm)	= 0.04 inches (A dime is 1 mm thick.)	
1 centimeter (cm)	= 10 millimeters (A slice of bread is about 1 cm thick.)	
	= 0.4 inches	
1 meter (m)	⇒ 100 centimeters (The maximum length of a baseball bat is about 1 m.)	
	= 3.28 feet	
	= 1.1 yards	
1 kilometer (km)	= 1,000 meters (Five average city blocks are about 1 km.)	
	= 0.62 miles	

Mass (measured in weight on Earth)

1 gram (g)	= .035 ounces (A potato chip weighs about 1 gr.)
1 kilogram (kg)	= 1,000 grams (This book weighs 1.8 kg.)
	= 2.2 pounds

Area

1 square centimeter (cm^2)	= 0.155 square inches (A dollar bill covers about 102 cm^2.)
1 square meter (m^2)	= 10.76 square feet (A standard table-tennis table covers about 4 m^2.)
	= 1.2 square yards

Astronomical distances

1 astronomical unit (AU)	= 149.6 million km (The average Earth-Sun distance)
	= 93 million miles
1 light-year	= 9.46 trillion km (The *distance* traveled by light in one Earth year)
	= 5.88 trillion miles
	= 63,240 AU
Speed of light	= 299,792.5 km per second
	= 186,282.4 miles per second

Temperature scale comparisons

At sea level ice melts at:	273 kelvins 0° Celsius 32° Fahrenheit	Human body temperature:	310 kelvins 37° Celsius 98.6° Fahrenheit
Room temperature:	293 kelvins 20° Celsius 68° Fahrenheit	At sea level water boils at:	373 kelvins 100° Celsius 212° Fahrenheit

Observing the Universe

The sites in this selective list represent U. S. and Canadian planetariums, observatories, and space or science museums open to the public or by appointment. Readers are encouraged to call local schools, museums, and astronomical societies to learn of additional places of interest.

If you plan to visit one of these sites, be sure to telephone first for fees and schedules.

Alabama
Robert R. Meyer Planetarium, Birmingham Southern College, Birmingham.
U. S. Space and Rocket Center, Huntsville.
W. A. Gayle Planetarium, Montgomery.

Arizona
Lowell Observatory, Flagstaff.
Museum of Astrogeology, Meteor Crater, Flagstaff.
Center for Meteorite Studies, Arizona State University, Tempe.
Kitt Peak National Observatory, Tucson.
Flandreau Planetarium, University of Arizona, Tucson.

Arkansas
University of Arkansas Planetarium, Little Rock.

California
California Museum of Science and Industry, Los Angeles.
Griffith Observatory, Los Angeles.
Big Bear Solar Observatory, California Institute of Technology, Big Bear.
Palomar Observatory, California Institute of Technology, North San Diego County.
Reuben H. Fleet Space Theater and Science Center, San Diego.
Alexander F. Morrison Planetarium, San Francisco.
Lick Observatory, San Jose.

Colorado
Charles C. Gates Planetarium, Denver Museum of Natural History, Denver.

Connecticut
The Discovery Museum, Inc., Planetarium, Bridgeport.
Gengras Planetarium, Science Center of Connecticut, Hartford.
Stamford Observatory and Edgerton Memorial Planetarium, Stamford Museum and Nature Center, Stamford.

Delaware
Mount Cuba Astronomical Observatory, Wilmington.

District of Columbia
National Air and Space Museum, Smithsonian Institution.
Rock Creek Nature Center.
United States Naval Observatory.

Florida
United States Air Force Space Museum, Cape Canaveral Air Force Station.
Alexander Brest Planetarium, Museum of Science and History, Jacksonville.
NASA-John F. Kennedy Space Center, Merritt Island.
The John Young Planetarium, Orlando Science Center, Orlando.
South Florida Science Museum, West Palm Beach.

Hawaii
Bishop Museum Planetarium, Honolulu.
Mauna Kea Observatory, Hilo.

Illinois
The Adler Planetarium, Chicago.
Museum of Science and Industry, Chicago.
Dearborn Observatory, Northwestern University, Evanston
John Deere Planetarium, Augustana College, Rock Island.

Indiana
Koch Planetarium, Evansville Museum of Arts and Sciences, Evansville.
J. I. Holcomb Observatory and Planetarium, Butler University, Indianapolis.

Iowa
Sanford Museum and Planetarium, Cherokee.
Sargent Space Theater, Science Center of Iowa, Des Moines.

Grout Museum of History and Science Planetarium, Waterloo.

Kansas
Clyde W. Tombaugh Observatory, University of Kansas, Lawrence.
Omnisphere and Science Center, Wichita.

Kentucky
Hardin Planetarium, Western Kentucky University, Bowling Green.
Joseph Rauch Memorial Planetarium, University of Louisville, Louisville.

Louisiana
Louisiana Arts and Science Center, Baton Rouge.
Lafayette Natural History Museum and Planetarium, Lafayette.

Maine
University of Maine, Maynard E. Jordan Observatory and Planetarium, Orono.
Southworth Planetarium, Portland.

Maryland
Davis Planetarium, Maryland Science Center, Baltimore.
University of Maryland Observatory, College Park.

Massachusetts
Boston Museum of Science, Boston.
The Center for Astrophysics: Harvard College Observatory and Smithsonian Astrophysical Observatory, Cambridge.
Maria Mitchell Observatories, Maria Mitchell Association, Nantucket.
Seymour Planetarium, Springfield Science Museum, Springfield.

Michigan
Cranbrook Institute of Science, Bloomfield Hills.
Detroit Science Center, Detroit.
Abrams Planetarium, Michigan State University, East Lansing.
Robert T. Longway Planetarium, Flint.

Minnesota
Minneapolis Planetarium, Minneapolis.
The Science Museum of Minnesota, St. Paul.

Mississippi
Russell C. Davis Planetarium, Jackson.

Missouri
Laws Observatory, University of Missouri, Columbia.
Kansas City Museum of History and Science Planetarium, Kansas City.
St. Louis Science Center, St. Louis.

Montana
Museum of the Rockies, Bozeman.

Nebraska
J. M. McDonald Planetarium, Hastings Museum, Hastings.
Ralph Mueller Planetarium, University of Nebraska State Museum, Lincoln.

Nevada
Fleischmann Planetarium, University of Nevada, Reno.

New Hampshire
Shattuck Observatory, Dartmouth College, Hanover.
Plymouth State College Planetarium, Plymouth.

New Jersey
Morris Museum, Morristown.
Newark Museum Planetarium, Newark.
New Jersey State Museum, Trenton.

New Mexico
Robert H. Goddard Planetarium, Roswell Museum and Art Center, Roswell.
National Solar Observatory, Sunspot.

New York
Kellogg Observatory, Buffalo Museum of Science, Buffalo.
Vanderbilt Planetarium, Centerport.
American Museum-Hayden Planetarium, New York.
Strasenburgh Planetarium, Rochester Museum and Science Center, Rochester.
Andrus Space Transit Planetarium, Hudson River Museum of Westchester, Yonkers.

North Carolina
Morehead Planetarium and Observatory, University of North Carolina, Chapel Hill.
Kelly Space Voyager Planetarium, Discovery Place, Charlotte.

North Dakota
Minot State College Observatory, Minot.

Ohio
Mueller Planetarium and Observatory, Cleveland Museum of Natural History, Cleveland.
Battelle Planetarium, Center of Science and Industry, Columbus.
Dayton Museum of Natural History, Dayton.
Neil Armstrong Air and Space Museum, Wapakoneta.

Oklahoma
Air Space Museum, Kirkpatrick Planetarium, Omniplex Science Museum, Oklahoma City.

Oregon
Pine Mountain Observatory, University of Oregon, Bend.
Murdock Sky Theater, Oregon Museum of Science and Industry, Portland.

Pennsylvania
State Museum of Pennsylvania Planetarium, Harrisburg.
Franklin Institute Science Museum and Planetarium, Philadelphia.
Buhl Planetarium, Carnegie Science Center, Pittsburgh.

South Carolina
Gibbes Planetarium, Columbia Museum of Art, Columbia.

Planetarium, Roper Mountain Science Center, Greenville.

Tennessee
Clarence T. Jones Observatory, University of Tennessee, Chattanooga.
Bays Mountain Park Nature Center, Planetarium, Kingsport.

Texas
Texas Memorial Museum, Austin.
The Southwest Museum of Science and Technology/Science Place Planetarium, Dallas.
McDonald Observatory at Mount Locke, Fort Davis.
Margaret Root Brown Observatory and Burke Baker Planetarium, Houston Museum of Natural Science, Houston.
Space Center, Houston.
Hanger 9, Edward H. White II Memorial Museum, Brooks Air Force Base, San Antonio.

Utah
Hansen Planetarium, Salt Lake City.

Vermont
Fairbanks Museum and Planetarium, St. Johnsbury.

Virginia
Leander McCormick Observatory, University of Virginia, Charlottesville.
Chesapeake Planetarium, Chesapeake.
Virginia Air & Space Center, Hampton.
Science Museum of Virginia, UNIVERSE Planetarium/Space Theater, Richmond.

Washington
Battelle Observatory (Rattlesnake Mountain Observatory), Richland.
Pacific Science Center, Seattle.

West Virginia
Sunrise Museum, Inc., Planetarium, Charleston.
The National Radio Astronomy Observatory, Green Bank.

Wisconsin
Washburn Observatory, University of Wisconsin, Madison.
Yerkes Observatory, University of Chicago, Williams Bay.

Wyoming
Wyoming Infrared Observatory, University of Wyoming, Laramie.

CANADA:
Alberta
The Science Center, Calgary.
Edmonton Space Science Center, Edmonton.

British Columbia
H. R. MacMillan Planetarium, Vancouver.
Dominion Astrophysical Observatory, Victoria.

Manitoba
Lockhart Planetarium, University of Manitoba, Winnipeg.
Manitoba Museum of Man and Nature Planetarium, Winnipeg.

Nova Scotia
Burke-Gaffney Observatory, Halifax.

Ontario
Cronyn Memorial Observatory, London.
National Museum of Science and Technology, Ottawa.
David Dunlap Observatory, Richmond Hill.
McLaughlin Planetarium, Toronto.
Ontario Science Centre, Toronto.

Quebec
Dow Planetarium, Montreal.

We'd like to thank...

It would have been impossible to put this book together without the help of our major consultants, listed on page 4, and many other people.

We are particularly indebted to the staff of NASA's Planetary Division, especially to Joseph M. Boyce, William E. Brunk, Angelo Guastaferro, Michael R. Helton, Gordon A. McKay, Rodney A. Mills, Robert E. Murphy, and William L. Quaide for their time and encouragement. Among others at NASA who assisted us are Donald L. DeVincenzi, Richard M. Farrel, Albert G. Opp, Erwin R. Schmerling, and Roger R. Williamson. We thank Paul D. Lowman, Jr., of the Goddard Space Flight Center; and Frank Bristow, Ellis D. Miner, Ray L. Newburn, Jr., Jurrie Van Der Woude, and Donald Yeomans of the Jet Propulsion Laboratory.

Donald M. Hunten and Bradford A. Smith of the University of Arizona gave us considerable help. We are also indebted to George O. Abell, Richard P. Binzel, A.G.W. Cameron, Dale P. Cruikshank, Stephen E. Dwornik, William K. Hartmann, James W. Head, Judith M. Hobart, Alfred S. McEwen, William C. Miller, Roman Smoluchowski, Laurence A. Soderblom, Peter B. Stifel, Jill Tarter, and George W. Wetherill for sharing their knowledge.

We received valuable assistance from the staff of the National Air and Space Museum, especially James D. Dean, Frederick C. Durant III, and Catherine D. Scott. We thank the staffs of the many observatories we consulted, and Roger Cayrel,

Pierre Bely, and Philippe Bourlon of the Canada-France-Hawaii Telescope. Connie S. Rodriguez at Kitt Peak National Observatory, NOAO, was especially generous with her time. Sally J. Bensusen, Robert S. Harrington, Paul M. Routly, P. Kenneth Seidelmann, and Thomas C. Van Flandern of the U. S. Naval Observatory also deserve special thanks.

Other people to whom we are grateful are: David C. Black, Don Campbell, Charles F. Capen, Clark Chapman, Charles C. Counselman, Raymond Davis, Jr., Amahl S. Drake, Stillman Drake, Diana Eck, John A. Eddy, Sidney W. Fox, Sharon Gibbs, Einar Haugen, Philip B. James, James G. Lawless, George Lovi, Brian Marsden, Brian Mason, Harold Masursky, Janet Mattei, Joanna A. McClellan, Jeffrey Meyerriecks, Derek Price, Frederick L. Scarf, and William A. Schopf. Dennis di Cicco and Dennis Milon of *Sky and Telescope* helped with photographs. The staff of our National Geographic Library was extremely helpful as always. We thank the girls and boys who acted as consultants on both text and illustrations.

Henning W. Leidecker computed data for the Lund Observatory map on pages 238-239, and for the asteroid and comet orbit diagrams on pages 144 and 208.

The editors also thank the Czechoslovakian publishers, Artia, for permission to reproduce the painting by Zdenek Burian on pages 102-103.

Index

Illustrations are in **boldface type (127).** Text references are in lightface type (56) and definitions in the text are in *italic type (218).*

Type composition by the Typographic section of National Geographic Production Services, Pre-Press Division. Color separations by Beck Engraving Co., Inc., Philadelphia, Pa.; The J. Wm. Reed Co., Alexandria, Va.; The Lanman Companies, Washington, D. C.; Penn Colour Graphics, Inc., Huntingdon Valley, Pa.; Progressive Color Corporation, Rockville, Md.; and Chanticleer Company, Inc., New York, N.Y. Printed and bound by R. R. Donnelley and Sons Co., Willard, Ohio. Paper by Consolidated/Alling & Cory, Willow Grove, Pa.

Illustration Credits

Symbols used in this list: NGP-National Geographic Photographer. NGS-National Geographic Staff. NGPA-National Geographic Publications Art. CIT-California Institute of Technology. JPL-Jet Propulsion Laboratory. NASA-National Aeronautics and Space Administration. NOAO-National Optical Astronomy Observatories.

Cover by John Berkey, Frank and Jeff Lavaty Associates. 1-3, Zeplin Productions Inc. 6-7, Howard Sochurek.

Beginnings
8-9, Peter Lloyd. 10-11, Peter Lloyd. 12-13, Thomas E. Hooper. 13, NGPA. 14-15, Peter Lloyd. 15 right, Ron Miller. 16-17, Peter Lloyd. 18, Charles and Ray Eames from Jagiellonian Library, Cracow, Poland. 19 top, Peter Lloyd. bottom, Ron Miller. 20, Charles and Ray Eames. 21, Peter Lloyd, after an illustration by Gordon Davies in *The Discovery of the Universe*, by Tony Osman, Usborne Publishing Ltd. 22, Peter Lloyd. 23 top left, Lick Observatory. top right, Owen Gingerich, courtesy of Houghton Library, Harvard University. bottom, Ron Miller. 24-25, Peter Lloyd. 26-27, James A. Sugar, Black Star. 27, Jean-Leon Huens. 28-29, Peter Lloyd. 30, Roger Ressmeyer, Starlight. 31, Ron Miller. 32-33, Jonathan Blair. 34, Ron Miller. 34-35, Roger Ressmeyer, Starlight.

The Sun & Planets
36-37, Peter Lloyd. 38, Davis Meltzer. 39 top and bottom left, NGPA. top and bottom right, Paul M. Breeden. 40-41, Ludek Pesek with additions by Davis Meltzer. 42 left, Paul M. Breeden. 42-43, Davis Meltzer. 43 center, NGPA. 44-45, Michael R. Whelan. 46-47, Michael R. Whelan.

The Sun
48-49, Ludek Pesek. 50 left, Michael R. Whelan. 50-51, Davis Meltzer. 51 top right, Metacolor. 52, William H. Regan and Maxwell T. Sandford, Los Alamos Scientific Laboratory. 53, Big Bear Solar Observatory, CIT. 54-55, NOAO. 56 top, Big Bear Solar Observatory, CIT. 56-57, NGPA. 58-59, Peter Lloyd. 60, Helmut K. Wimmer. 61, Ira Block. 62-63, Ira Block. 64-65, Ira Block. 65 top, Ron Miller. 66-67, photos for composite by Naval Research Lab., American Science and Engineering, High Altitude Observatory (all NASA), and NOAO.

Mercury
68-69, Ludek Pesek. 70 top left, Michael R. Whelan. bottom left, NGPA. right, Davis Meltzer. 71, Davis Meltzer. 72-73, Chris Foss. 74 left, JPL, NASA. right, Pierre Mion. 75, NASA. bottom, Pierre Mion. 76-77, Ludek Pesek. 79, Jay L. Inge.

Venus
80-81, Ludek Pesek. 82 top left, Michael R. Whelan. bottom left, NGPA. right, Davis Meltzer. 83, Davis Meltzer. 84, Ludek Pesek. 85, Tass from Sovfoto. 86-87, Ludek Pesek. 88 left, NGPA. right, Ames Research Center, NASA. 90, Davis Meltzer. 91, Jay L. Inge.

Earth
92-93, Ludek Pesek. 94 top left, Michael R. Whelan. bottom left, NGPA. right, Davis Meltzer. 95, Davis Meltzer. 96, Jaime Quintero. 97, NASA. 98-99, Ken Sakamoto, Black Star. 99, Jaime Quintero. 100-101, Ludek Pesek. 102-103, Zdenek Burian. 104-105, NASA. 106-107, William H. Bond, NGS. 108-109, Loren McIntyre. 110-111, Jay L. Inge. 112, Pierre Mion. 113, Scott Bolden, NGS. 114 top left, Michael R. Whelan. 114-115, Davis Meltzer. 116-117 top, Lunar Orbiter 5, NASA. bottom, NASA. 118-119, Eugene A. Cernan, NASA. 120-121, Jay L. Inge. 122-123, NASA.

Mars
124-125, Ludek Pesek. 126 top left, Michael R. Whelan. bottom left, NGPA. right, Davis Meltzer. 127, Davis Meltzer. top right, NGPA. 128-129, Vincent Di Fate. 130-131, NASA. 132-133, Ludek Pesek. 134, NGPA and NASA. 134-135, Graphic Films, JPL, NASA. 136, JPL, NASA. 137, NASA. 138-139, Jay L. Inge. 140-141, NASA.

Asteroids & Meteors
142-143, Ludek Pesek. 144, NGPA. top left, Michael R. Whelan. 145, Davis Meltzer. 146-147, Chris Foss. 147, JPL, NASA. 148, Mr. and Mrs. James M. Baker. 149 top, Albert Moldvay. bottom, Robert T. Dodd. 150-151, William K. Hartmann. 151, NGPA.

Jupiter
152-153, Ludek Pesek. 154 top left, Michael R. Whelan. bottom left, NGPA. right, Davis Meltzer. 155, Davis Meltzer. 156-157, NASA. 158, Davis Meltzer. 159, Vincent Di Fate. 160-161, Ludek Pesek. 162-163, Jay L. Inge. 164, NASA. 165, NGPA. 166-167, NASA. 169, NASA. 170-171, Ludek Pesek.

Saturn
172-173, Ludek Pesek. 174 top left, Michael R. Whelan. bottom left, NGPA. right, Davis Meltzer. 175, Davis Meltzer. 176, NASA. 176-177, Ludek Pesek. 178-179, NASA. 180-181, NGPA. 181 right, NASA. 182-183, Lloyd K. Townsend. 183 bottom, NASA. 184-185, Ludek Pesek.

Uranus
186-187, Ludek Pesek. 188 top left, Michael R. Whelan. bottom left, NGPA. right, Davis Meltzer. 189, Davis Meltzer. 190, Jean-Leon Huens. 191, JPL, NASA. 192, Digital Cartography by Kathleen Edwards, Kevin Mullins, and Christopher Isbell, USGS.

Neptune
194-195, Ludek Pesek. 196 top left, Michael R. Whelan. bottom left, NGPA. right, Davis Meltzer. 197, Davis Meltzer. 198, JPL, NASA. 199, Ludek Pesek.

Pluto
200-201, Ludek Pesek. 202 top left, Michael R. Whelan. bottom left, NGPA. right, Davis Meltzer. 203, Davis Meltzer. center left, U. S. Naval Observatory. bottom, NGPA. 204, Lowell Observatory. 205, Ludek Pesek.

Comets
206-207, Ludek Pesek. 208 left, Michael R. Whelan. right, NGPA. 209 top, Davis Meltzer. bottom, NGPA. 210, Max-Planck-Institut Für Aeronomie. 211, Victor R. Boswell, Jr., NGP. 212-213, Helmut K. Wimmer. 213, NASA. 214-215, Chris Foss. 215, NGPA. 216-217, Chris Foss.

Deep Space
218-219, Helmut K. Wimmer. 220-221, Jaime Quintero. 222, Michael R. Whelan. 223, photo composite by Ira Block. 224 top, Helmut K. Wimmer. bottom, NGPA. 225, James A. Sugar, Black Star. 226 top, NOAO. 226-227, Pierre Mion. 227, Mark Seidler. 228, "Beta Lyrae" by Chesley Bonestell, courtesy of Mr. and Mrs. Thomas J. Mullen, Jr. 229, Helmut K. Wimmer. 230-231, Ludek Pesek. 232, David Malin © Royal Observatory, Edinburgh. 233 upper, Palomar Observatory, CIT. lower, NOAO. 234, Helmut K. Wimmer. 235 top, Palomar Observatory, CIT. bottom, Lick Observatory. 236, Helmut K. Wimmer. 237, Victor J. Kelley. 238-239, Lund Observatory. 239, Helmut K. Wimmer. 240, Palomar Observatory, CIT. 241 top left, Anglo-Australian Telescope Board, David Malin. top right, Palomar Observatory, CIT. bottom left, NOAO. bottom right, Hans Vehrenberg, Hansen Planetarium. 242-243, Helmut K. Wimmer. 244 top, Arnold H. Rots and Whitney W. Shane, National Radio Astronomy Observatory and The Netherlands Foundation for Radio Astronomy. 244-245, Ron Miller. 246-247 top, NOAO. bottom, from "Violent Tides between Galaxies" by Alar and Juri Toomre, copyright December 1973 by Scientific American, Inc. 248, Bradford A. Smith, University of Arizona, and Richard J. Terrile, JPL. 249, Roger Ressmeyer, Starlight. 250, James A. Sugar, Black Star. 251 left, National Astronomy and Ionosphere Center, Cornell University. right, Robert W. Madden, NGP. 252-253, Vincent Di Fate.

Shuttles & Starships
254-255, Vincent Di Fate. 256-257, Jon Schneeberger, Ted Johnson, Jr., Lawrence B. Maurer, and Anthony Peritore, NGS. 258 bottom, NASA. 258-259, from the IMAX®/OMNIMAX® film "The Dream is Alive" © Smithsonian Institution and Lockheed Corporation. 260-261, Boeing Aerospace Company: Paul Hudson. 262-263, Sydney Mead. 264-265, Vincent Di Fate. 266-267, "Lunar Farside Research Base" by Pierre Mion, courtesy National Air and Space Museum, Smithsonian Institution. 267, "Mining an Asteroid" by Chesley Bonestell, courtesy of Ames Research Center, NASA. 268-269, Sydney Mead. 270-271, Sydney Mead.

Highlights
272 left, Archives, CIT. top center, Edwin L. Wisherd. bottom center, NASA. right, courtesy of Bell Laboratories. 273 left, U. S. Army, Official. top center, Sovfoto. bottom center, Novosti Press Agency. top right, Igor Snegirev, Novosti Press Agency. bottom right, NASA. 274 top left, Flip Schulke, Black Star. bottom left, BBC. top center, Novosti Press Agency. bottom center, Maarten Schmidt, CIT. top right, JPL, NASA. bottom right, Douglas S. Chaffee. 275 top left, James A. McDivitt, NASA. bottom left, Thomas P. Stafford, NASA. top right, Bruce Dale, NGP for NASA. bottom right, Jonathan Blair. 276 top left, Neil A. Armstrong, NASA. bottom left, A. Patnesky, NASA. top center, Eugene A. Cernan, NASA. 276-277, NASA. 277 top left, Carl and Linda Sagan, Frank Drake, National Astronomy and Ionosphere Center, Cornell University. center, Skylab 4, NASA. right, NASA.

Library of Congress CIP Data
Gallant, Roy A.
National Geographic picture atlas of our universe.
Includes index.
Summary: Text, photographs, paintings, and maps explore the history of astronomy, the solar system, the universe, and new space discoveries.
1. Astronomy—Pictorial works. 2. Astronomy—Charts, diagrams, etc. [1. Astronomy]
I. Sedeen, Margaret. II. Newhouse, Elizabeth L. III. National Geographic Book Service. IV. Title. V. Title: Our universe.
QB68.G34   1986   523   86-8775
ISBN 0-87044-644-4 (alk. paper)
ISBN 0-87044-645-2 (lib. bdg. : alk. paper)

1994 revision
ISBN 0-7922-2731-X reg.
ISBN 0-7922-2732-8 deluxe